Comprehensive

INJECTION MOULDING

INJECTION MOULDING

INJECTION MOULDING

Other CBS book by the same author

Batra: Design of Blow Moulds

Comprehensive

INJECTION MOULDING

R.C. Batra B Tech (Hons)
German Advisor (Retd)

CBS Publishers & Distributors Pvt. Ltd.

New Delhi • Bengaluru • Chennai • Kochi • Kolkata • Mumbai
Hyderabad • Nagpur • Patna • Pune • Vijayawada

ISBN: 978-81-239-1879-2 (PB)
ISBN: 978-81-239-1940-9 (HB)

First Edition: 2011
Reprint: 2011, 2018

Published by **Satish Kumar Jain** and produced by **Varun Jain** for
CBS Publishers & Distributors Pvt. Ltd.,
4819/XI Prahlad Street, 24 Ansari Road, Daryaganj, New Delhi - 110002
delhi@cbspd.com, cbspubs@airtelmail.in • www.cbspd.com
Ph.: 23289259, 23266861, 23266867 • Fax: 011-23243014

Corporate Office: 204 FIE, Industrial Area, Patparganj, Delhi - 110 092
Ph: 49344934 • Fax: 011-49344935
E-mail: publishing@cbspd.com • publicity@cbspd.com

Branches:
• *Bengaluru:* 2975, 17th Cross, K.R. Road, Bansankari 2nd Stage,
  Bengaluru - 70 • Ph: +91-80-26771678/79 • Fax: +91-80-26771680
  E-mail: cbsbng@gmail.com, bangalore@cbspd.com
• *Chennai:* No. 7, Subbaraya Street, Shenoy Nagar, Chennai - 600030
  Ph: +91-44-26681266, 26680620 • Fax: +91-44-42032115
  E-mail: chennai@cbspd.com
• *Kochi:* Ashana House, 39/1904, A.M. Thomas Road, Valanjambalam,
  Ernakulum, Kochi • Ph: +91-484-4059061-65
  Fax: +91-484-4059065 • E-mail: cochin@cbspd.com
• *Kolkata:* 6-B, Ground Floor, Rameshwar Shaw Road, Kolkata - 700014
  Ph: +91-33-22891126/7/8 • E-mail: kolkata@cbspd.com
• *Mumbai:* 83-C, Dr. E. Moses Road, Worli, Mumbai - 400018
  Ph: +91-9833017933, 022-24902340/41 • E-mail: mumbai@cbspd.com

Representatives:

• Hyderabad: 0-9885175004	• Nagpur: 0-9021734563
• Patna: 0-9334159340	• Pune: 0-9623451994
• Jharkhand: 0-9811541605	• Uttarakhand: 0-9716462459

Printed at:
India Binding House, Noida, UP (India)

Introduction

Among all the industrial manufacturing processes, injection moulding is probably the only one which, starting with the raw material, delivers a finished product in the final shape, colour and finish in one shot. Initially developed for moulding of thermoplastics, it has made forays into many alien fields hitherto unthought of and has brought accuracy, efficiency and economy to them. The scope of the injection moulding process has widened much beyond the limited field of thermoplastics and has contributed enormously towards bringing most modern appliances and comforts within the reach of everyone. The trend in development of materials and equipment indicates that the injection moulding process would overshadow all other related manufacturing methods.

Injection moulding constitutes the main theme of the book which aims at explaining the process of conventional injection moulding in all practical details because it forms the basis for all other variations. Special attention has been paid to the rules of mould design which are a key to successful processing. These guidelines are also applicable to the tooling for other related processes directly or indirectly. The description of the equipment has been handled as much as necessary to understand the process.

The guideline, while compiling the book, has been the practical application of the contents. A large number of illustrations, diagrams and tables should help the reader in understanding and application of the knowledge collected from widely diverse sources like plastics processors, manufacturers of raw materials, machinery and moulds, research theses, journals, books, seminars, workshops, lectures, exhibitions, experience gathered by the author in the German plastics industry and advice of many well wishers.

RC Batra

Acknowledgements

Compiling a book covering over a dozen, widely diverse versions of the injection moulding process has been an ambitious venture which would not have been possible without the active support of numerous organisations, predominantly German, which generously provided me literature, illustrations and advice, and permitted me for free use of the contents.

Arburgs	Figs 2.19, 2.28 to 2.30, 12.1 to 12.6, 12.12, 12.15
BASF	Figs 1.4, 1.16, 1.17, 2.3, 4.24, 4.67
Dr Boy	Fig. 2.31
Dow Corning	Fig. 15.1
Engel	Fig. 8.4, Table 8.1
EOS	Figs 5.6, 5.7
Ewikon	Fig. 4.92
Guenther	Fig. 14.8
HASCO	Figs 4.37, 4.94
Heitech	Figs 4.81 to 4.83, 4.87 to 4.89
HS-Heitzelemente	Fig. 4.79
IKV/Sulzer	Figs 11.1, 11.2
Kunststoff-Institut, Luedenscheid	Figs 3.4 to 3.7
Schoenthaler Dr W	Figs 2.26, 14.1
Stamixco	Figs 2.11, 2.12
Strack-Normalien	Fig. 4.53
Vogel-Buchverlag	Figs 1.1, 1.3, 1.11 to 1.13, 1.18, 1.19, 16.1, 16.2

Wittmann-Battenfeld Figs 12.7, 12.10, 12.11, 12.18 to 12.20
Xintech Figs 4.78, 4.84, 4.86

I must point out that the source of some information, especially the older one, was impossible to trace as the data had been acquired in seminars, lectures, technical fairs or from journals. Likewise, publishers of some older books did not exist any more and could not be contacted. I am nevertheless equally thankful to all known and unknown donors.

I do not wish to miss expressing my sincere gratitude to my learned friends Mr AP Sharma, Mr Goswami, Mr Satya Prakash and Prof Chaturvedi among many others who have helped me to realise this project. Among the students who lent a helping hand, Ishant deserves a special mention.

RC Batra

to

Mr TG Punwani
The unforgettable pioneer

Contents

1

Plastics

- History of plastics
- Formation of plastics
- Types of plastics
- The molecular structure and properties
- Common plastics and thermoplastics
- Classification of their properties
- The form of raw material
- Preparation of plastics for processing
- Recycling
- Acronyms

HISTORY

Plastics is a generic term applying to a wide range of materials with long chains of molecules. While solid in the finished state, at some stage during their manufacturing, plastics are soft enough to be formed into various shapes, most commonly through the application, either singly or in combination, of heat and pressure. They may be natural or be produced either by modification of some natural material by chemical means or by synthesis of elements such as carbon, hydrogen, nitrogen, oxygen, etc. into compounds of long chains.

Mankind has known and used natural plastics since time immemorial. Vedas, the ancient holy books of Hindus, dwell upon shellac, a natural resin from a tree and its applications. Also Mahabharata, the great epic of Aryans, mentions its employment for building of a palace which was later set on fire to kill the occupants. More recently, that is in 1596, John Hyglen undertook a scientific mission to India on behest of the king of Portugal to investigate this material unknown to the west. He has left descriptions about its various applications being practiced by Indians in those days.

1

Five centuries ago, Columbus saw Red Indians playing with lumps of rubber. Later, in 1731, La Codamine observed South American natives using rubber for waterproofing of fabrics and shoes. It has been recorded that the king of Portugal sent his shoes with the next expedition to be waterproofed. The properties of natural rubber, that is latex, were later improved by treating it with sulphur and it acquired worldwide importance when automobiles became a major means of transport.

In the middle of the seventeenth century, the English traveller John Tradescant observed Malayens heating up a tree resin and forming knife handles out of it. Gutta percha, as the material was called, proved of great value to the west as a coating for electrical cables and held sway in this application till polyethylene was developed during the second world war.

Till 1860, the only plastics known were those obtained from nature, viz. wax, shellac, gutta percha and rubber. Only the latter had been modified so that it could also be turned to a hard substance called ebonite. However, they had whetted the appetite and the hunt for more had begun. The search for a substitute for silk proved another impetus for experimentation. Efforts to treat natural cellulose with chemicals bore fruit and cellulose nitrate was born. It yielded silk-like yarn but was highly inflammable. The fabric woven from the new material could catch fire from its own static charge. However, further developments turned it into a mouldable resin. Celluloid became synonymous with thermoplastics and was moulded indiscriminately into all sorts of utility articles. Celluloid toys became the rage of the period, but some other applications failed and brought bad name to "plastics".

A fully synthetic material, developed with phenol and formaldehyde, started the era of plastics in the first decade of the twentieth century. It was the first man-made synthetic material and its uses were still to be found. The inventor, Leo Hendrik Baekland, christened it Bakelite and it almost became another name for all thermosetting plastics. Principally, it found application in electrical field but was also used for articles like buttons, grips, handles, crockery, telephone sets, car parts, etc.

Urea Formaldehyde, which followed soon after, was also a thermosetting resin distinguished by its transparency.

Two thermoplastics, viz. cellulose acetate and polyvinyl chloride, which made their debut in 1927, paved the path for future plastics. The fact that the process of injection moulding had also achieved mass production maturity around that time, helped their exploitation in many fields.

The decade 1930–40 was most eventful for synthetic materials. Many new plastics were introduced, which helped conquer fields hitherto untouched. Acrylics, developed in 1936, proved an excellent substitute for glass in automobiles and aeroplanes. The most sensational find was nylon, initially developed with an aim to provide a better substitute for silk for ladies' stockings, but was found to be an ideal material for demanding technical applications. The third member of the formaldehyde family, melamine formaldehyde, complemented the range of thermosets in 1939. It brought unbreakable crockery in fancy colours within reach of the common man.

The development of synthetic materials was accelerated during the second world war with the aim to find substitutes for natural materials needed for military equipment. The accidental discovery of polyethylene in England in 1942 had its share in tilting the balance in favour of allies. It helped development of the radar which made the most dangerous weapon of Germany, viz. the undetectable submarines, detectable and vulnerable. Another plastics used extensively in the military hardware was acrylics in place of glass in aeroplanes. Cockpits too were thermoformed out of this tough thermoplastics.

The development set in motion by the war did not lose its momentum after armistice and the materials discovered and tried during that period were employed in civilian applications. More and more synthetic materials appeared on the scene and helped fill up many gaps besides accelerating development in other fields using plastics.

Although the development of new plastics is continuing unabated, the procedure of the search has undergone a diametric change. In the initial stages, plastics was developed (and in some cases stumbled upon accidentally) first and then examined for its properties and utility. Now the research takes place with a definite set of properties and applications in mind as one has learnt how to manipulate the macromolecules.

It had been prophesied by a visionary long ago that plastics could one day furnish all materials needed by the mankind with the exception of cutting edges and materials which have to withstand direct application of heat. That day does not seem to be far.

Table 1.1 lists some of the more common plastics, processed by injection moulding in order of their appearance.

Table 1.1: Introduction of plastics	
Year	*Plastics*
1868	Cellulose nitrate
1909	Phenol formaldehyde
1918	Urea formaldehyde
1927	Cellulose acetate
1927	Polyvinyl chloride
1935	Melamine formaldehyde
1935	Polyamide 6 6
1936	Polymethyl methacrylate
1936	Polyurethane
1936	Polystyrene
1939	Polyamide 6
1942	Polyester
1942	Low density polyethylene
1943	Silicones
1945	Cellulose propionate
1947	Epoxide
1948	Acrylonitrile butadiene styrene
1955	High density polyethylene
1956	Polyactal
1957	Polypropylene
1957	Polycarbonate
1964	Polyphenylene oxide
1964	Polyimide
1964	Ethylene vinyl acetate
1965	Polysulfone
1971	Polyphenylene sulfone
1972	Thermoplastic elastomer
1972	Polyether sulfone

contd...

contd...

1975	Liquid Crystal Polymer
1980	Polyether Ether Ketone
1982	Polyetheramide

The popularity of plastics may be attributed to the following factors:

- They are ideal for mass production
- They can be transparent, opaque and coloured
- They do not rust.
- They are light and tough
- They are good insulators against heat and electricity
- They can be used with foodstuff
- They are resistant to chemicals
- Their properties can be tailored for particular applications
- They can be processed in the desired colour and finish right away.
- Their surface finish can be matched with the non-plastics parts they are assembled with.
- Their surface can be decorated in numerous ways.
- Multi-component articles can be produced in one operation
- Most of them can be recycled

FORMATION OF PLASTICS

Till middle of the last century, coal had been the main source of the synthetic materials labelled collectively as plastics. The ingredients of the first fully synthetic material, viz. phenol and formaldehyde were by-products of the destructive distillation of coal. The derivatives of the by-products of coal also formed the basis for many other plastics like polystyrene and polyamides. Reaction of coke with calcium oxide yields calcium carbide which produces acetylene. Acetylene, in turn, is the starting point for production of acrylonitrile, vinyl chloride, vinyl acetate and other vinyl monomers.

The development of the petrochemical industry and that of plastics have gone hand in hand after the second world war. The research work carried on petroleum during the war also benefited plastics as more direct routes were found for mass production of these synthetic materials.

It is worth mentioning that research work is going on to find alternative, renewable sources of plastics. Not only some plant constituents such as proteins, starch, vegetable oils and cellulose but also some animal products like wool, skins, leather, gelatine, tallow, feathers, whey and casein may also become potential sources of plastics. Starch and cellulose are the naturally occurring polymers. Natural fibres like jute, flax, straw, hemp, etc. consist mainly of cellulose which, in turn, has already been modified to mouldable plastics like cellulose acetate, propionate and butyriate. Residuals from agricultural crops such as straw, stems, hulls, bran, husks have the same chemical composition as the wood fibre and form a cheap and abundant source of cellulose. The process has started and plastics have been produced on commercial scale out of corn. Plastics from agricultural sources can reduce dependence on fossil fuels and help reduce carbon dioxide emission. Vegetable oils and sugar too may become future sources of plastics. It must, however, be pointed out that natural sources have a lower proportion of carbohydrates than the crude oil and require more energy for their transformation into plastics in contrast to the fossil oil.

Fractional Distillation of Oil

The crude oil contains hydrocarbons in various combinations like gases, light and heavy oils, petrol, coal tar, etc. These are separated by fractional distillation (Fig. 1.1). Heavy benzene or naphtha is the raw material for plastics. By application of heat (850°C) in presence of catalysts, its molecule is cracked into smaller compounds of carbon and hydrogen like ethylene and ethane (Fig. 1.2).

Polymerisation

Ethylene, a gas, is separated by liquefying the mixture. It forms basis for many plastics as well as for other compounds. C_2H_4 is its monomer or one building block. By the process of polymerisation, which is formation of a long chain or a so-called macromolecule consisting of many individual building blocks or monomers, it is converted into polyethylene, which is a solid polymer.

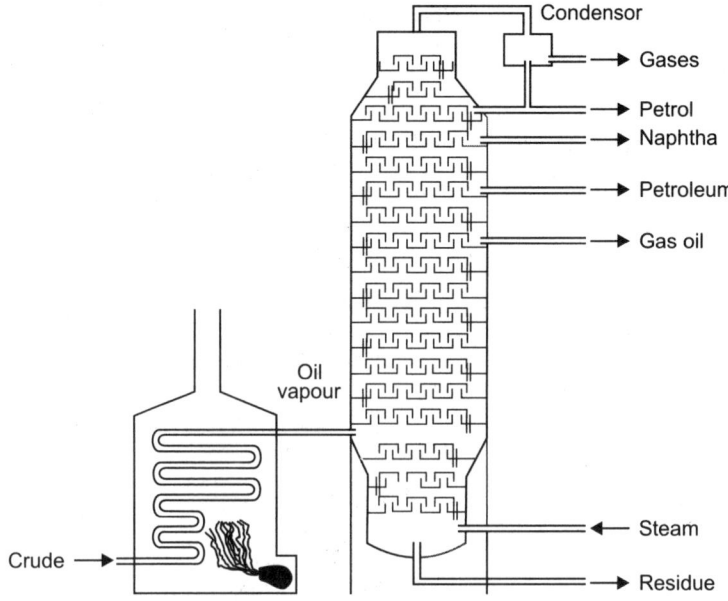

Fig. 1.1

Fig. 1.2

Polymerisation can be represented graphically by the following diagram (Fig. 1.3):

Figure 1.4 shows a molecule of polyethylene consisting of a number of monomers, each having two carbon atoms (black) and four hydrogen atoms (white).

As mentioned before, the monomer of ethylene forms the model for many other plastics. For example, the monomer of propylene is similar to that of ethylene with the difference that one of the four hydrogen atom S is replaced by a CH_3 group. Through the process of polymerisation, we get the plastics known as polypropylene (Fig. 1.5).

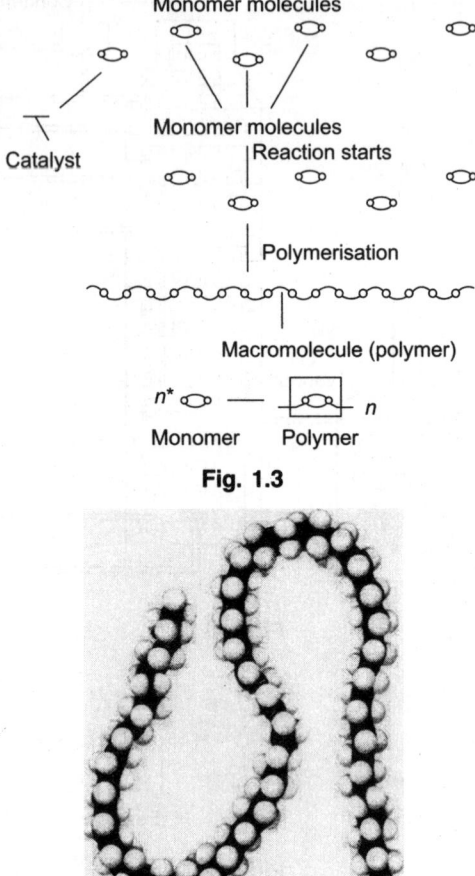

Monomer molecules

Catalyst

Monomer molecules
Reaction starts

Polymerisation

Macromolecule (polymer)

$n^* \bigcirc\!\!\!-\!\!\!\bigcirc \quad - \quad \boxed{\bigcirc\!\!\!-\!\!\!\bigcirc} \; n$

Monomer   Polymer

Fig. 1.3

Fig. 1.4

One atom of chlorine and three of hydrogen in combination with one atom of carbon form the monomer of vinyl chloride (Fig. 1.6).

```
  H         H
  |         |
  C   =     C
  |         |
  H     H—C—H
          |
          H
```

Fig. 1.5

```
  H    H
  |    |
  C  = C
  |    |
  H    CL
```

Fig. 1.6

The monomer of acrylonitrile differs from that of ethylene in that the former contains a group of CN in the place of one atom of hydrogen (Fig. 1.7).

Similarly, a benzol ring in place of one hydrogen atom represents a molecule of styrene (Fig. 1.8).

The monomer of tetrafluoroethylene has four atoms of flourine joined with two atoms of carbon (Fig. 1.9).

| **Fig. 1.7** | **Fig. 1.8** | **Fig. 1.9** |

Polymerisation is not limited to formation of macromolecules out of one type of monomers only. The process can be carried out with two or more monomers such as with acrylonitrile and styrene to yield SAN, a copolymer (Fig. 1.10) or with acrylonitrile, styrene and butadiene ABS, a terpolymer.

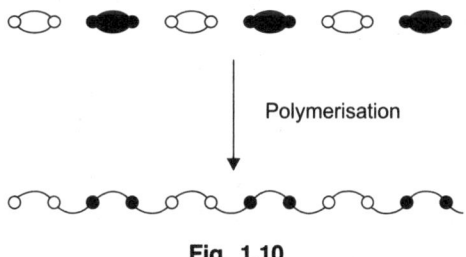

Fig. 1.10

Polycondensation

Polycondensation besides polymerisation, is another process of forming plastics out of two different types of building blocks. Unlike polymerisation, this process sets a by-product, mostly H_2O free. Nylons (polyamides) are formed by this process. The polycondensation of a polyamide molecule from diamine and dicarboxylic acid is represented by the Fig. 1.11.

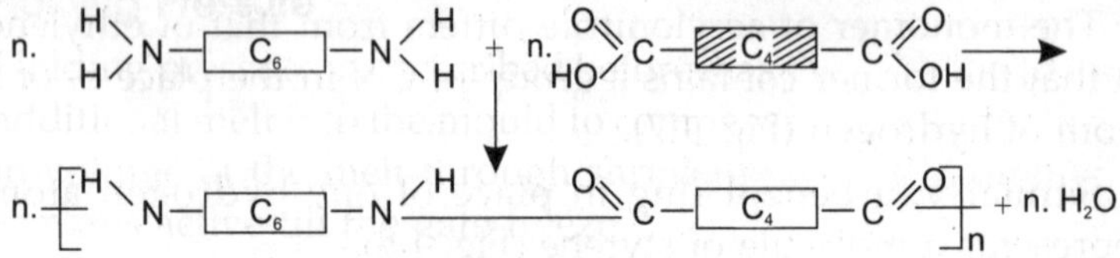

Fig. 1.11

The nomenclature for different polyamides, such as PA 6 6, PA 6 10, etc. is derived from the number of carbon atoms in each part of the combined monomer. For example, PA 6 10 consists of two groups, viz. $-NH(CH_2)_6 NH-$ and $-CO(CH_2)_8 CO-$

If three or more groups are joined together by polycondensation, the final product is a thermosetting plastics. Its molecule is no more linear as with a thermoplastics but three dimensional.

Polyaddition

Polyaddition is the process of forming a new product out of two different molecular building blocks without giving rise to a by-product. Polyurethanes are produced by polyaddition of isocynate and dialcohol. The process differs from polymerisation in that atoms are exchanged between the two monomers as shown in Fig. 1.12.

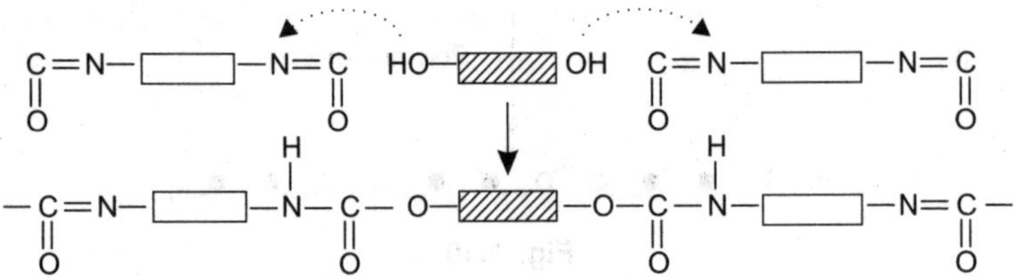

Fig. 1.12

TYPES OF PLASTICS

Plastics are divided into two basic categories:
a. Thermosetting plastics
b. Thermoplastics

Thermosetting Plastics

Thermosetting plastics "set" with the heat permanently into a shape which cannot be changed by reheating. Their molecules undergo an irreversible chemical change and become cross linked once for all with the first heat to a definite level. Thermosetting plastics have three-dimensional macro-molecules. The binding force among the molecules is the chemical valency, which is strong (Fig. 1.13). This is the reason, why the thermosetting plastics are hard and do not flow under heat and pressure after processing.

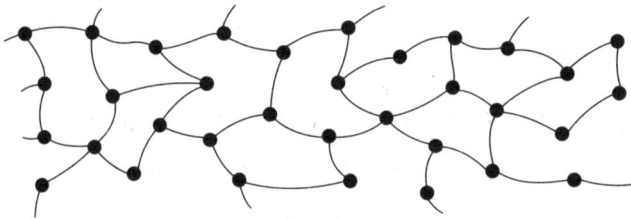

Fig. 1.13

Elastomers

Elastomers constitute a sub-class of the thermosets with reversible elastic properties. Their molecules are partially cross linked and the polymers retain a degree of elasticity (Fig. 1.14). Elastomers are rubbery at room temperature and regain their shape after the deforming force is removed. Like thermosets, elastomers too cannot be re-formed after first reaction. Liquid silicone rubbers belong to this category.

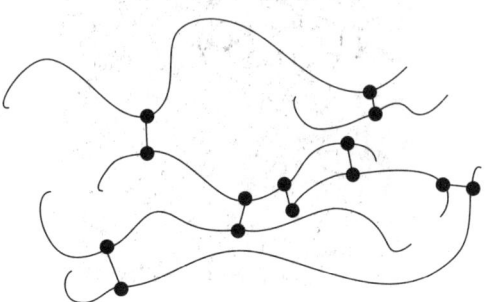

Fig. 1.14

Thermoplastics

Thermoplastics soften and subsequently melt at certain temperature and can be formed to an intended shape. On cooling, they become solid again but the process of softening with heat and re-shaping can be repeated because the heat does not effect any chemical change in their structure. Thermoplastics have linear macromolecules consisting of monomers, which are bound together by the strong force of their valencies. However, the holding force between the macromolecules is the coordinate bond or an electrostatic force, which is weak and gets weaker under temperature and pressure (Fig. 1.15).

Based on the molecular structure, thermoplastics have two distinct sub-classes:

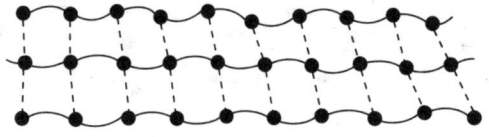

Fig. 1.15

Amorphous Thermoplastics

The macromolecules of plastics belonging to this class are not arranged in an orderly fashion (Fig. 1.16). When melted, their volume increases up to 15%.

Fig. 1.16

Their linear shrinkage is relatively small and varies between 0.3-0.8%.
- Amorphous thermoplastics are rigid and hard.
- They are transparent in their basic form, with a few exceptions.
- Their specific gravity ranges from 1.05 to 1.4
- Amorphous thermoplastics do not have a distinct melting point. They soften at some stage, called glass transition temperature, when heated. Their heat content is relatively low.
- The most common amorphous thermoplastics are PS, SAN, PMMA, PC, etc.

Semi-crystalline Thermoplastics

Semi-crystalline thermoplastics have part of their molecular chains arranged in a parallel mode. The rest is disorderly like those of the amorphous plastics (Fig. 1.17). Upon melting, all molecular chains become disorderly.
- Their volume increases by about 30 % or more upon melting.
- Their linear shrinkage is relatively large; it varies between 1.5–3.5 %
- In their natural state, they are translucent to opaque.
- Their specific gravity lies between 0.9 and 1.3
- While melting, they require latent heat. They have a distinct melting point. Their heat content is relatively high.
- The most common semi-crystalline thermoplastics are: LDPE, HDPE, PP, PA, POM, PET, etc.
  There are a few special forms of thermoplastics like:

Fig. 1.17

Liquid Crystal Polymers

Liquid crystal polymers, though crystalline, form a special class of thermoplastics. They possess rod like crystals, which retain their form even in the molten state. LC Polymers have properties, which make them an ideal replacement of metals in many cases. Their shrinkage is low and even negative along the direction of flow for some compounds.

Polyblends

Polyblends are a mixture of two or more thermoplastics having different properties. The resultant blend has properties which are average of those of its constituents. The blends are created with following aims:
- Improving polymer processability
- Supplementing the physical and mechanical properties
- Developing tailor-made materials for specific applications
- Reducing the material cost by mixing a low cost material with an expensive one
- Recycling the mixed scrap

As most of the blends are immiscible, phase separation occurs during processing. In order to overcome this difficulty, special coupling agents including some copolymers called compatibilizers are added to the mixture.

Thermoplastic Elastomers

In contrast to the thermosetting elastomers, the thermoplastics elastomers owe their elasticity to a combination of hard and soft phases. Butadiene, isoprene, butylene, etc. form the soft phase whereas styrene acts as the hard component. The possibilities of combinations are unlimited.

The thermoplastics elastomers are inferior in their elastics properties but possess the advantage of recycling, typical of thermoplastics.

Biologically Degradable Plastics

Biologically degradable plastics are polymers which can be disintegrated into natural ingredients like water, carbon dioxide, methane, minerals, etc. through the action of natural agents like enzymes, bacteria, fungus, algae, etc. Some plastics can be disintegrated by natural light, hydrolysis or oxidation.

Additives

Polymers can rarely be processed in their pure form. In order to create or improve the processability and to modify some characteristics or to incorporate some missing properties, certain substances, usually chemicals but also inorganic compounds, are added to them. They do not alter the basic chemical character of the plastics. Thermosetting polymers are rarely processed in their pure form; they may contain fillers to the extent of 50–75% by weight. The most common additives are:

- **Plasticisers** lend flexibility and softness to otherwise hard and brittle polymers (PVC is one such polymer). The plasticisers also enhance the impact strength.
- **Lubricants** help reduce viscosity by creating a film between the melt and the plasticising equipment as well as between the melt and the die/mould. Thus the thermally sensitive materials do not get overheated. The internal lubricants facilitate easy gliding of molecules past one another with the same effect.
- **Antioxidants** prevent or minimise the weakening or damaging effect of high processing temperatures.
- **Heat stabilisers** protect plastics against high temperatures during processing. PVC, which is thermally sensitive and unstable, cannot be moulded without these additives.
- **Colorants** add the desired colour to the plastics and also provide protection against light. Some luminous and fluorescent pigments lend visibility to plastics in the dark. The colorants may be of organic or inorganic origin. Some of them are harmful and are not allowed in contact with foodstuff.
- **Flame retardants** make plastics self-extinguishing.
- **UV-Stabilisers** provide protection against the damaging effect of ultraviolet rays.
- **Blowing agents** turn plastics into foams.
- **Compatibilizers** help make alloys or composites out of different plastics.

- **Fillers** or extenders are added primarily to increase the volume and reduce the material cost. They also modify certain properties like tensile strength, modulus of elasticity, sound absorption, shrinkage behaviour, etc. The most common fillers are carbon black, calcium carbonate, talcum, clay, glass fibres, mica, wood flour, etc. Except wood, all fillers are heavier than the plastics.

- **Accelerators** are mostly used with thermosets to accelerate curing, i.e. cross linking of molecular chains, at elevated temperatures.

- **Inhibitors** perform the function of retarding the process of cross linking of molecular chains at lower temperatures. They prolong the shelf life of thermosetting resins. During injection moulding of thermosets, they act against premature curing of the polymer in the plasticising unit.

- **Antistatic agents** are added to make plastics electrically conductive. Carbon black is the most common additive. Other agents to reduce the electrical resistance are carbon fibres, stainless steel fibres, polyanin powder, etc.

Reinforced Plastics

To enhance certain properties of plastics, reinforcing components are physically mixed into them. Glass fibres are the most common addition. The other reinforcing ingredients are glass beads, mineral fibres, talcum, mica, etc. For special applications, steel fibres have also been incorporated. The reinforcing components show beneficial effect in case of many physical properties but there are some adverse effects also.

There is an increase in the values of:
Modulus of elasticity, creep resistance, tensile and flexural strength, heat conductivity, high temperature resistance, specific gravity and impact strength below the freezing point.

Following attributes experience a decrease:
Coefficient of thermal expansion, cooling time as the minerals have lower specific heat and better heat conductivity, mould shrinkage, impact strength at room temperature, flowability and weld line strength.

THE MOLECULAR STRUCTURE AND PROPERTIES

Synthetic materials, called plastics collectively, are built of long chains of molecules consisting of monomers having atoms of elements like carbon, hydrogen, oxygen, nitrogen, silicone, etc. These long chains of molecules are named macromolecules and may contain thousands of monomers. This is why, the plastics are referred to as polymers. The length of the chains is not constant.

The physical and mechanical properties of plastics depend upon the shape and length of the macromolecules as well as on forces which hold them together. Some knowledge of their structure is helpful in understanding their basic properties and behaviour.

Thermoplastics have linear macromolecules, sometimes with side branches, consisting of monomers bound by the strong force of their valencies. The holding force between the macromolecules is the co-ordinate bond or an electrostatic force (Fig. 1.15), which is weak and gets weaker with increase in distance, temperature and pressure. The linear structure permits the polymer segments to get closer to each other than the branched structure so that the former is denser (Fig. 1.18). The branched side chains, on the other hand, will keep the main polymer backbones further apart from each other (Fig. 1.19). The longer the chain, the greater is the effect. The denser polymer would have higher specific gravity and higher rigidity.

The two varieties of polyethylene, which have the same chemical formula, are a specific example of this effect. The molecules of HDPE are linear without side branches; they lie closer together and their co-ordinate forces are more effective. LDPE on the other hand has macromolecules with side

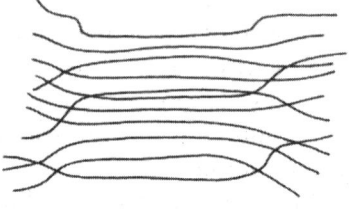

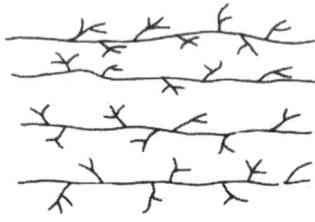

Fig. 1.18　　　　　　**Fig. 1.19**

branches; the molecules cannot be as close to one another as in the case of HDPE. The co-ordinate forces are therefore weaker and the plastics is less rigid.

Out of the same reason, the polymer with branched chains and more space between adjacent chains would have higher permeability to gases.

Linear materials with closely packed chains will have higher intermolecular forces and therefore higher tensile strength.

Likewise the linear materials, being closely packed, will be stiffer because of less room for bending of the backbones. Their modulus of elasticity will be higher.

As the molecular forces binding a high density material are higher, it takes a higher amount of heat energy to separate the linear configuration. Consequently, the heat distortion temperature of materials with linear macromolecules will be higher.

Hardness, which can also be defined as resistance to penetration, will be higher for materials having closely packed molecular chains.

Creep resistance or the resistance to distorting forces too, is higher with materials of linear molecular chains because of higher intermolecular forces.

Flowability is lower in case of stronger molecular attraction as the flow is opposed by the cohesive forces.

Compressibility: Branched chains have space between them. Hence these materials are more easily compressible than the ones with straight chains.

Impact strength is, however, better with branched materials, which are more flexible.

Crystalline thermoplastics have a part of their molecules lying closer together in an orderly fashion. Their physical and mechanical properties are closely linked not only with their crystalline structure but also with the extent and form of crystallinity. The compactness as a result of closeness results in better cohesion between the molecular chains, which increases the density, the stiffness, the heat resistance, etc. The effect is similar to that of the unbranched chains. It may be summed up that:

- The more the crystallinity, the higher the density.
- The more the crystallinity, the greater the stiffness as the closely packed crystals have less space to rotate.
- The more the crystallinity, the higher the tensile strength as more force is needed to break the intimate bonds.
- The more the crystallinity, the more the hardness (resistance to penetration) because of the closely packed molecular chains.
- The more the crystallinity, the less the permeability to gases as there are less gaps between the chains.
- The more the crystallinity, the higher the softening temperature as more heat energy is needed to overcome the cohesive forces.
- The more the crystallinity, the more the volume expansion on melting as the freed compact chains get disorderly and occupy more space.
- The more the crystallinity, the greater the shrinkage on cooling. Ordered molecules occupy less space.
- The more the crystallinity, the less the impact strength. Crystals propagate cracks.
- The more the crystallinity, the less the resistance to stress cracking.
- Difference in crystallinity in different sections results in more warpage.

The extent of crystallinity in a moulding out of semi-crystalline thermoplastics depends upon factors like mould temperature, cycle time, injection and holding pressure, rate of cooling, etc. Slow cooling results in higher degree of crystallisation.

Flowability: Plastics melts are visco-elastic fluids. The characteristic of polymers, most relevant to the process of injection moulding, is their flowability or the viscosity at moulding temperatures. It is the inner resistance of the melt to flow. The viscosity of a plastics decides the force needed to plasticise and transport the melt, the extent of pressure to fill the mould cavity, the time for filling and the force needed to prevent the mould from opening under the filling pressure. In other words, it influences the design of the moulding machine as well as that of the mould and the moulding.

The flow process of plastics melt involves shear, which is caused by adherence of the melt to the adjacent surfaces of the plasticising unit and the walls of the mould cavity. The force needed to make it move introduces shear and the rate of shear is proportional to the applied force. Shear is also caused by the resistance of successive layers of the melt gliding past one another.

Plastics are non-newtonian fluids. Their viscosity decreases with the increasing rate of shear. In other words, they flow more easily with higher speed. The viscosity of a particular plastics is therefore, not an absolute factor. It depends upon the rate of shear. The temperature, too, has an influence on the rate of flow. As the temperature of the polymer rises, the plastics expands and the distance between the macromolecules increases which facilitates their movement past one another. Consequently, less force is needed to make the melt flow. The length of the macromolecular chains has a direct influence on the flowability. The shorter chains can glide past one another more easily than the longer ones. This is why, the material grades with longer chains or higher molecular weights are more viscous. The grades meant for injection moulding should have lower viscosity.

Viscosity can be measured and expressed in several ways. The method more commonly employed in practice is that of measuring melt flow index.

The melt flow index (MFI) is a measure of apparent dynamic fluidity or flowability of thermoplastics. It is the weight of the melt in grams, extruded through a standard die at specific temperature under a specific force in 10 minutes in a standard plastmeter (Fig. 1.20). The MFI is expressed as a number. A higher number denotes better flowability, which indicates lower average molecular weight. The MFI number, however, holds good only for individual thermoplastics. Melt Flow Indices of different plastics cannot be compared with one another.

The melt flow index number is not an absolute measure of actual viscosity but it is a practical scale to compare the flowability and suitability of different material grade for a particular application. For instance, to mould a thin-walled component like a disposable injection syringe, a grade with a

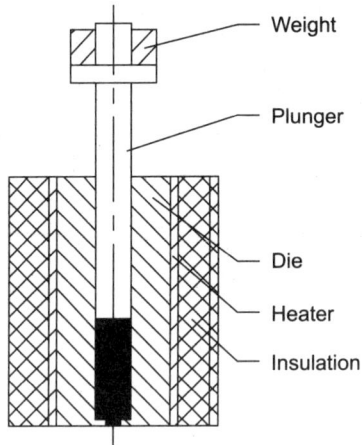

Fig. 1.20

higher MFI will be more suitable than the one with lower MFI, being employed for making a sturdy household container.

The method of measuring the melt flow index also delivers useful information about the uniformity of the raw material being produced and delivered. It is employed by the raw material producers as well as by the buyers to check instantly the conformance of the material.

Flowability of heat sensitive materials like PVC cannot be measured by the above process. The measure of their flowability is the so-called K value.

The viscosity of a thermally sensitive thermoplastics is measured by dissolving it in a solvent and determining the speed of flow of the solution in a capillary viscometer relative to that of the pure solvent.

Relative viscosity = time taken by solution/time taken by the solvent

Knowing values of the relative viscosity and of the concentration of the solution, the K-value of the polymer can be read out of a corresponding table. A higher K value indicates higher viscosity or lower flowability.

The solvent used for PVC for determining the relative viscosity is cyclohexanon.

Molecular weight is another specification which gives information about many characteristics of the polymer. It is the sum of the atomic weight of all atoms constituting one macromolecule of a thermoplastics. All chains in a polymer are, however, not of the same length. Therefore, a particular material grade is known by its average molecular weight. The grades with higher molecular weight have longer macromolecules and hence lower flowability but they have properties characteristic of longer molecular chains such as higher tensile and impact strength, better creep resistance and chemical resistance due to increase in entanglement and intermolecular attraction between the adjacent chains.

COMMON PLASTICS AND THEIR PROPERTIES

Thermosetting Plastics

It may be stressed upon in the outset that the thermosetting plastics are rarely moulded in their pure form. They are always mixed with additives like glass, cotton, and organic fibres, talcum, wood flour, cotton fabric, etc. Their properties depend upon the type and extent of additives.

Phenol Formaldehyde

Phenol formaldehyde has not only the distinction of being the first fully synthetic material; it also became a synonym for the thermosets under the name of Bakelite. In his search of a synthetic substitute for shellac, Mr. Baekland, an immigrant to USA, experimented with phenol and formaldehyde and developed, in 1907, a brownish substance which he baptised as Bakelite. It was found that it could harden irreversibly under the influence of heat and pressure and the reinforced versions offered unlimited possibilities of its use. The reinforced phenol formaldehyde has:

- High strength, stiffness and hardness.
- Low cold flow.
- High heat resistance.
- High resistance to organic solvents, neutral chemicals, weak acids and alkalis (it is attacked by the strong acids and alkalis).
- Resistance to stress cracking.
- Good electrical insulation.

Phenol formaldehyde burns with a bright sooty flame but extinguishes automatically outside the source of ignition, emitting a smell of phenol and formaldehyde and, depending upon the reinforcement, of ammonia too. It is not suitable for use with foodstuff. It is brownish in its natural form but gets darker in light. Hence it is available only in dark colours.

Following substances are used as reinforcing ingredients, either singly or in combinations, for different applications:

- Wood flour.
- Wood pulp.
- Cotton fibres/cotton fabric chips.
- Synthetic fibres.
- Asbestos fibres.
- Stone powder.
- Mica.
- Rubber.

Specific gravity	1.4–1.9
injection unit temperature	60–85°C
Mould temperature	160–190°C
Service temperature without load	
Short time	130–150°C
Long time	110–130°C
Mould shrinkage	0–1.0%
Post shrinkage	0–0.5%

(All values depend upon the type and amount of reinforcement).

Conversion processes: Compression, transfer and injection moulding, extrusion.

Uses: Electrical connectors, switches, meter boxes, spools, handles, knobs, ash trays, components for pumps, screw caps, gears, etc.

Urea Formaldehyde (UF)

The search for a transparent synthetic material resulted in the development of urea formaldehyde a decade after the creation of phenol formaldehyde. To maintain the transparency, the urea formaldehyde is primarily reinforced with bleached cellulose. However, other ingredients such as stone powder, fibres, asbestos, etc. are also added when transparency is not required. Reinforced UF has the following properties:

- High mechanical strength, stiffness and surface hardness.
- Very good electrical properties.

- High surface gloss.
- Proneness to stress cracking.
- Poor dimensional stability because of variable water absorption.
- Higher shrinkage than phenol formaldchyde.

Urea formaldehyde burns with a yellow flame and is self-extinguishing. It is sensitive to moisture. It is not suitable for applications involving food articles.

Urea Formaldehyde can be injection moulded in special formulations containing agents which retard its curing at moulding temperatures.

Specific gravity	1.5
Moulding unit temperature	70–80°C
Mould temperature	160–170°C
Service temperature without load	
Short term	100°C
Long term	70°C
Mould shrinkage	0.5–0.8%
Post shrinkage	0.7–1.2%

Conversion processes: Compression moulding, transfer moulding, injection moulding.

Uses: Electrical parts like plugs, sockets, meter boxes, lamps, screw caps and closures for cosmetic containers, knobs, toilet seats, furniture parts, etc.

Melamine Formaldehyde (MF)

Reinforced with organic and inorganic substances, the melamine formaldehyde resins have the following properties:
- High surface hardness and scratch resistance
- High surface lustre
- Good heat resistance
- Resistance to moisture
- High post-shrinkage leading to cracks
- Usable with foodstuff
- Prone to stress cracking
- Unsuitable for use with boiling water.

Specific gravity	1.5–1.9
Mould temperature	150–170°C
Service temperature without load	

	Short term	120–170°C
	Long term	80–120°C
Mould shrinkage		0.1–0.9%
Post shrinkage		0.3–1.2%

Conversion processes: Compression moulding, transfer moulding, injection moulding.

Uses: Almost same as for UF but due to better creep strength and admissibility for use in contact with eatables, it is used for crockery, electrical parts needing creep resistance, housings for gadgets, furniture parts etc.

Special note: Combinations of PF and MF are employed to get:
- Good dielectric properties
- Better creep resistance
- Less shrinkage and better dimensional stability
- Lower tendency to form cracks under moisture and heat
- Lower cost

Apart from some special formulations in pourable powders, melamine requires auxiliary equipment for feeding with the injection moulding machine.

Unsaturated Polyester Resin

Usually referred to without the adjective, the polyester resin is transparent in its natural form but also brittle. This is why, it is always used with reinforcing ingredients, preferably with glass fibres. It also burns readily with a bright, sooty flame, emitting a sweetish smell. Hence fire retardant are added. Reinforced polyester resins possess following characteristics:
- High strength, stiffness and hardness
- High heat resistance
- Very good electrical properties
- Very little water absorption
- Good weathering stability

Specific gravity		1.8–2.0
Mould temperature		160–170°C
Service temperature without load		
	Short term	200°C
	Long term	150–160°C
Mould shrinkage		0.2–0.7%
Post shrinkage		0–0.1%

Conversion processes: Compression moulding, transfer moulding, injection moulding.

Uses: Electrical components like plugs, sockets, fuse boxes, insulation material, spools, lights, housings for electrical tools, car parts like spark plug holder, ignition distributor, ignition coil etc.

Epoxide Resins

The excellent flowability of epoxide resins makes them ideal for the moulding of intricate components. Epoxy resins have very good electrical and chemical properties. The usual reinforcement of glass fibres has a positive influence on mechanical properties and heat stability.

The epoxy resins get darker in colour with the time and are therefore pigmented in dark tones. They are combustible and hence need addition of fire retardants when used for electrical components. Because of their very good flowability and very low shrinkage, they are highly suitable for:

• Thin walled components
• Components with close tolerances.
• Components with inserts. Epoxy resins adhere well to metals.
• Components with inserts. Epoxy resins have low risk of stress cracking.

Specific gravity	1.8–1.9
Mould temperature	125–175°C
Service temperature without load	
Short term	180°C
Long term	130°C
Mould shrinkage	0.2–0.3%
Post shrinkage	0–0.1%

Conversion processes: Compression and transfer moulding, injection moulding, casting, etc.

Uses: Multi-pole connectors, plugs and sockets, condensers, resisters, winding spools, diodes, transistors, relays, etc.

Thermoplastics

Amorphous Thermoplastics

Cellulose acetate (CA), cellulose propionate and cellulose acetate butyriate form a family of thermoplastics from the group of plastics derived from natural resources. CA is still produced and used in sizeable quantities. It is based on natural cellulose, a polysaccharide ($C_6H_{10}O_5$) which forms the main ingredient of all plants.

- CA has high toughness, strength and resilience.
- It is transparent but can be coloured in all hues and shades.
- It has a brilliant surface.
- It is highly scratch resistant.
- It is resistant to stress cracking.
- It is resistant to organic solvents, mineral oils, benzene, fats, etc.
- It burns with a green-yellowish flame and emits a pungent smell like burning paper.
- It is not suitable for outdoor applications.
- It absorbs water and needs pre-drying before processing. Dimensions are effected by absorbed water.

Special attribute: It takes very little electrostatic charge and does not attract dust.

Specific gravity		1.3
Moulding temperature		180–220°C
Service temperature without load		
	Short term	0–80°C
	Long term	0–70°C
Mould shrinkage		0.4–0.7%

Conversion processes: Injection moulding, extrusion, blow moulding. thermoforming.

Uses: Telephone sets, microphone housing, tool handles and grips, hammer heads, buttons, keys, spectacle frames, pens, toys, combs and brushes, shoe heels, etc.

Polystyrene (PS), also referred to as the General Purpose Polystyrene, is an amorphous thermoplastics with a shiny surface. It is transparent in natural state. It is hard and brittle. The wear resistance is average. As it absorbs water only to a negligible extent, its dimensional stability is excellent.

- It burns with a sooty flame and smells sweet.
- It is resistant to weak acids and alkalis but swells in gasoline, benzene and other organic solvents.
- It is not impervious to steam, oxygen, carbon dioxide, etc.
- It is prone to stress cracking.

- It emits a metallic sound when dropped on hard surface.

Specific gravity	1.05
Moulding temperature	170–280°C
Working temperature without load	
Short term	80–85°C
Long term	65–75°C
Mould shrinkage	0.4–0.6%

Conversion processes: Injection moulding, injection blow moulding, extrusion, thermoforming.

Uses: Packaging for food, medicine, cosmetics, disposable cutlery, disposable tumblers, transparent covers, drawing instruments, etc.

The following copolymers of polystyrene overcome various deficiencies like brittleness of polystyrene, retaining other attributes.

High Impact Polystyrene (HIP)

The GP Polystyrene modified with butadiene becomes tough and opaque but loses much of its shine. The wear resistance is average. It absorbs some water and needs pre-drying before processing.

- It burns with a sooty flame.
- It is resistant to stress cracking.
- It is resistant to weak acids and alkalis. It is attacked by organic solvents.

Specific gravity	1.05
Moulding temperature	190–280°C
Working temperature without load	
Short term	60–70°C
Long term	50–60°C
Mould shrinkage	0.4–0.7%

Conversion processes: Injection moulding, blow moulding, extrusion.

Uses: Packaging for food, medicines, cosmetics, housing for small gadgets, machinery, components for household machines, fans, parts for furniture, toys, etc.

Acrylonitrile Styrene (SAN)

This copolymer of styrene retains the shine and transparency of the parent material. It is tough and hard and has better heat resistance.

- It absorbs more water and needs pre-drying before processing.
- It burns like GP polystyrene.
- It is resistant to weak acid and alkalis. It is dissolved by organic solvents.
- It is resistant to stress cracking.
- It has better heat resistance than polystyrene.

Specific gravity	1.08
Moulding temperature	200–260°C
Service temperature without load	
Short term	95°C
Long term	85°C
Mould shrinkage	0.4–0.7%

Conversion processes: Injection moulding, blow moulding, extrusion, thermoforming.

Uses: Packaging for food, medicines, cosmetics, parts of household gadgets, telephone sets, knobs, keys and buttons, covers of car lights, traffic signals, pens, garden furniture, boat bodies, etc.

Acrylonitrile Butadiene Styrene (ABS), has better hardness, scratch resistance, toughness, heat resistance and resistance to chemicals than the other derivatives of styrene.

- It is yellowish opaque in natural state. A special formulation is transparent.
- It absorbs more water and must be pre-dried before processing.
- It is resistant to stress cracking.
- It burns with a sooty flame.

Special attribute. ABS can be metalised.

Specific gravity	1.04–1.06
Moulding temperature	200–260°C
Service temperature without load	
Short term	90–100°C
Long term	80–85°C
Mould shrinkage	0.4–0.7%

Conversion processes: Injection moulding, blow moulding, extrusion, calendering, thermoforming.

Uses: Housings for office and household machines, metalised radiator grills, dashboard components in automobiles, entertainment electronics, suitcases and boxes, safety helmets, knobs, keys, buttons etc.

Unplasticised polyvinyl chloride (uPVC) or rigid polyvinyl chloride is one of the most widely used plastics. Basically, it is an unstable polymer but a broad palette of stabilisers, additives and copolymers have made it the most prolific of synthetic materials so that it offers formulations for extremely diverse applications.

- It is a transparent thermoplastics with high mechanical strength, stiffness and hardness.
- It is highly resistant to chemicals.
- It is a good insulator for low voltages.
- It has a narrow temperature range for processing.
- It is resistant to stress cracking.
- It can be used for outdoor applications.
- It burns with a yellowish-orange pungent flame emitting chlorine but is self-extinguishing.

Specific gravity		1.4
Moulding temperature		180–210°C
Service temperature without load		
	Short term	75°C
	Long term	65°C
Mould shrinkage		0.4–0.5%

Conversion processes: Injection moulding, extrusion, blow moulding, calendering.

Uses: Pipes and fittings, medical disposables, transparent bottles and containers, profiles and sections for windows and furniture, vessels and fittings for the chemical industry, gramophone records, separators for batteries, etc.

Plasticised Polyvinyl Chloride (PVC-soft), is made flexible with the help of the chemicals called softeners and also with copolymers. Although most of the properties of the parent resin, viz. the uPVC are retained, the softeners bestow a degree of flexibility, which is dependent upon the amount of additives. Certain properties such as resistance to chemicals do get effected as the softeners may leach out with the passage of time or under adverse conditions. The flexibility may decrease and the resin may become brittle. More softener makes PVC more

susceptible to attack by chemicals. Like uPVC, the plasticised PVC is:
- Transparent
- Good insulator
- Self extinguishing by moderate amount of softeners
- Resistant to strong chemicals
- Brittle at sub-zero temperatures

Some of the additives may be injurious to health. Only selected additives are permitted for PVC-formulations meant for use as packaging of foodstuff.

Specific gravity	1.2–1.35
Moulding temperature	170–210°C
Service temperature without load	
Short term	~ 70°C
Long term	~ 50°C
Mould shrinkage	1.5–2.5%

Conversion processes: Injection moulding, extrusion, calendering, dip coating, rotational moulding.

Uses: Flexible pipes and tubes, floor tiles, gaskets and seals, curtains and table cloths, rain coats, artificial leather, straps and bands, cable coating, decorative foils, packaging films, medical tubing, medical disposables, shoes and shoe soles, tool grips, etc.

Polymethyl Methacrylate (PMMA), Also popularly known as acrylics, PMMA is an amorphous thermoplastics with high stiffness and hardness. It is fairly scratch resistant. Its surface can be polished to a high brilliance.
- It is transparent like water but copolymers have a yellowish tinge.
- It burns with a crackling flame and a fruity smell.
- It is prone to stress cracking.
- It is resistant to weak acids and alkalis.
- It is attacked by polar organic solvents.
- It is resistant to UV-rays; can be used outdoors.

Specific gravity	1.18
Moulding temperature	190–290°C
Service temperature without load	
Short term	90°C
Long term	75°C
Mould shrinkage	0.4–0.8%

Conversion processes: Injection moulding, extrusion.

Uses: Geometrical instruments, watch glasses, lenses, skylight covers, window panes, car tail light covers and reflectors, sanitary fittings, sign boards, costume jewellery, combs and brushes, handles for cutlery, etc.

Polycarbonate (PC), is an engineering thermoplastics, which exhibits very high stiffness, strength and hardness between extreme temperatures of −150° and +135°C.

- It is transparent like glass.
- It is a very good electrical insulator.
- It has very high heat resistance.
- It is resistant to weak acids and alkalis.
- It can be used in contact with foodstuff.
- It burns with a sooty, pungent smell of phenol but is self-extinguishing.
- It is prone to stress cracking.
- It absorbs water and necessitates pre-drying before processing.

Specific gravity	1.2
Moulding temperature	270–320°C
Service temperature without load	
Short term	−135 +160°C
Long term	−135 +135°C
Mould shrinkage	0.6–0.8%

Conversion processes: Injection moulding, extrusion, blow moulding.

Uses: Electrical and electronics components, Safety covers for electrical installations, winding spools, lamp covers, machine parts, valves, camera and projector housings, lenses, office machines' bodies and components, computer parts, traffic signals, headlight reflectors, car light covers, air conditioner parts, medical devices and disposable instruments, compact discs, etc.

Polyphenylene oxide (PPO), is an amorphous thermoplastics with a very high stiffness and hardness. Its heat resistance is excellent.

- It is transparent in natural state.
- It is highly resistant to chemicals.

- It is highly resistant to hydrolysis.
- It is self extinguishing.
- It absorbs very little water.

  Special attribute: It degenerates rapidly in air above 100°C. It can be metalised.

Specific gravity	1.06
Moulding temperature	260–310°C
Service temperature without load	
Short term	200°C
Long term	100°C
Mould shrinkage	0.4–0.7%

Conversion processes: Injection moulding, extrusion, blow moulding.

Uses: Electronic components, medical technical gadgets, components for washing machines, etc. foils for cable insulation, condensers, drawing paper and drawing foils, photographic film.

Note: Modified PPO as a polyblend with PS is produced by General Electric under the name of Noryl. It has all properties of PPO without the disadvantage of degeneration in the air. Its processability is superior to that of PPO. It is more commonly used than the unblended polymer.

Uses: Components for radio and TV, electrical parts, spools, electrical water heater, components for dish washing and clothes washing machines, hair dryers, sanitary fittings, water meter housing, office machines' parts, cameras, projectors, chromium plated parts for cars, grills, etc.

Polysulphone (PSU), is an engineering thermoplastics with very high stiffness, hardness and mechanical strength.

- It is highly resistant to chemicals.
- It is transparent like glass with a light yellowish tinge.
- It is highly heat resistant.
- It burns with bright smoky flame but is self-extinguishing.
- It is prone to stress cracking when in contact with certain chemicals.
- It is hygroscopic. It must be pre-dried before processing.

Specific gravity	1.24
Moulding temperature	320–380°C

Service temperature without load

	Short term	200°C
	Long term	150–170°C
Mould shrinkage		0.7–0.8%

Conversion processes: Injection moulding, extrusion, blow moulding, thermoforming.

Uses: Components of household and industrial machines, water heaters, automobiles, aeroplanes, projectors, electrical connectors, battery housing, foils for printed circuits, etc.

Special Note: Another polysulfone, viz. polyarylsulfone (PAS) has similar properties but can withstand temperatures up to 180°C.

Polyethyleneterephthalate (PETP), is a linear thermoplastics polyester of the semi-crystalline variety and as such it is white opaque. The share of the crystalline structure is about 30–40% but it can be reduced to such an extent with the help of co-monomers that the resultant polymer is transparent. Its remarkable properties are:

- Excellent hardness, stiffness
- Very good toughness, also at sub-zero temperatures.
- Good ageing characteristics. Low coefficient of friction.
- Good electrical insulation.
- Resistance to stress cracking.
- No reaction with water at room temperature, dilute acids, alcohols, oils and fats, etc.
- Reaction with high temperature steam, alkalis, oxidising acids, and organic solvents.
- Fire resistant, but burns with a sooty flame. Can be used with foodstuff.
- Absorbs very little water but needs pre-drying.

(Data for amorphous PETP)

Specific Gravity		1.33
Moulding temperature		260–300° C
Service Temperature without load		
	Short term	180°C
	Long term	100°C
Mould shrinkage		0.2–0.4 %(1.2–2.0% for linear PETP)

Conversion processes: Injection moulding, blow moulding, extrusion, thermoforming.

Uses: Moulded parts such as rollers, gears, wheels, bearings, guides, components for pumps, office machines, valves, telephones, etc.

Transparent bottles for aerated beverages, household fluids films and foils, monofilaments, foils for audio and video cassettes and electrical insulation, artificial lawns, garments etc.

PC + ABS-blend (e.g. Bayblend), The polyblend of poly-carbonate and acrylonitrile butadien styrene is amorphous in structure and combines the characteristics of the two polymers. The niveau of its properties is much higher than that of ABS but lower than of PC. It has:

- High impact and notch strength, even at sub-zero temperatures
- High stiffness
- High temperature resistance
- High dimensional precision
- Low distortion
- Low shrinkage
- Stability to UV-rays

Specific gravity	1.1–1.2
Moulding temperature	230–300°C
Service temperature without load	
Short term	112°C
Long term	95°C
Mould shrinkage	0.55–0.75%

Uses: Automobile components, electrical and electronic gadgets, household appliances, hobby and sports goods, etc.

Semi-crystalline Thermoplastics

Low density Polyethylene (LDPE)

Polyethylenes of different densities along with the polypropylene constitute the group of polyolefines. One common attribute of these semi-crystalline thermoplastics is their specific gravity, which is less than one so that they all float in water. The monomer ethylene is the simplest one, having two atoms of carbon and four of hydrogen (C_2H_4), which is also the basic building block of wax.

There are two distinct forms of polyethylene; the low density and the high density polyethylene. Most of their properties are similar; the difference lies in the extent.

LDPE is translucent to whitish-opaque, depending upon the thickness. It is flexible. Its molecules can arrange themselves parallel to one another in crystalline structures among the mass of randomly scattered molecules, while cooling from molten state to the solid phase. The degree of crystallisation may reach 40–50% depending upon the rate of cooling and the linearity of the molecules. The crystalline zones contribute stiffness. Its macromolecules are mostly branched.

- It possesses very high toughness.
- It has excellent electrical insulation properties.
- It absorbs very little water.
- It has very good resistance to all chemicals at room temperature.
- LD polyethylenes with higher molecular weight are more resistant to stress cracking.
- It burns without soot and smells like wax.

Specific gravity	0.91–0.93
Moulding temperature	160–270°C
Service temperature without load	

	Short term	80–90°C
	Long term	60–70°C
Mould shrinkage		1.5–3%

- The cross-linking varieties of LDPE, which become permanently set after treatment, have higher working temperatures.

Conversion processes: Injection moulding, extrusion, blow moulding, film blowing, rotational moulding.

Uses: Packaging for foodstuff and diverse articles as shrink film, bags, tubes and pipes, insulation for electrical cables, lids and closures for containers, sealing rings and gaskets, household articles, containers for chemical industry, etc.

High Density Polyethylene (HDPE), has a molecular structure, which is more linear than that of LDPE. It is also stiffer than the LDPE.

In natural state, HDPE is whitish in colour. Very thin layers may appear translucent.

HDPE has poor scratch resistant. Ultra high molecular high density versions possess very good surface hardness and sliding properties.

- It is resistant to almost all chemicals, except to strong acids, at room temperature.
- It is an excellent insulator to electricity.
- It burns with a bright, sootless flame, emitting smell of wax.
- It is prone to stress cracking.

Specific gravity	0.94–0.97
Moulding temperature	200–300° C
Service temperature without load	
Short term	90–100°C
Long term	70–80°C
Mould shrinkage	1.5–4.0%

Conversion processes: Injection moulding, extrusion, blow moulding.

Uses: Household articles, containers for food and general merchandise, petrol tanks and storage tanks for oils and chemicals, crates and containers for transport, bottles and closures, toys, films for packaging and insulation, etc.

The ultra high molecular varieties are used in textile industry for rollers and wheels, sliding chutes, implants in human body etc.

Polypropylene (PP), is the lightest member of the polyolefin family.

- It has high toughness, hardness and the tensile strength.
- Its scratch resistance is highest among polyolefins.
- It has high heat resistance. It can withstand boiling water.
- It has very good chemical resistance.
- It is translucent to opaque; some special formulations are transparent in thin sections.
- It burns with a soot-free flame and smells like wax.
- It becomes brittle at sub-zero temperatures.
- It is sensitive to UV-rays.

Note: With the help of nucleating agents, the pattern of crystallinity can be altered, in that instead of a few big clusters of crystals, many smaller clusters are formed. This structure results in less dispersion of light, conferring on the material more transparency. The nucleated polypropylene shrinks more but also more uniformly. The warpage is also reduced.

Special attribute: Thin, integral hinge out of PP is practically indestructible.

Specific gravity	0.9
Moulding temperature	220–300°C
Service temperature without load	
Short term	140°C
Long term	100°C
Mould shrinkage	1.0–2.5%

Conversion processes: Injection moulding, extrusion, blow moulding.

Uses: Pipes, tubes, profiles, sheets, foils, insulation for cables, components for electrical, electronic, automobile and chemical industry, sanitary fittings, household articles, medical equipment, transport boxes and crates, packaging for instruments and small machines like drills, car parts, toys, bags and sacks, packaging bands, carpeting, artificial lawns.

Polyamides

The specific aim of development of polyamides was the artificial yarn. Patented in 1937, PA 6 6 was introduced in USA in 1940 in the form of Nylon stockings. In Germany, simultaneous research led to the development of PA 6 and PA 6 6 for injection moulding and extrusion. They were found to be eminently suitable for engineering applications because of the following properties:

- Very high strength, stiffness and hardness.
- Very good heat endurance.
- Excellent wear resistance, good sliding characteristics, low friction.
- Very good shock absorption and damping.
- Resistance to solvents, gasoline, oils and fats.

Special attribute: All polyamides absorb moisture, which influences their properties. The toughness increases considerably but dimensions also increase and the electrical insulation decreases.

Table 1.2: Polyamides					
Attributes	*PA 6*	*PA 6 6*	*PA 6 10*	*PA 11*	*PA 12*
Specific gravity	1.13	1.14	1.08	1.0	1.02
Moulding temperature°C	230–290	260–320	230–290	200–270	200–270
Service temp. without load					
Short term °C	140–180	170–200	140–180	140–150	140–150
Long term °C	80–100	80–120	80–110	70–80	70–80
Mould shrinkage %	0.8–2.1	1.0–2.2	0.5–2.8	0.5–1.5	0.4–0.6
Moisture intake %	2.5–3.5	2.5–3.1	1.2–1.6	0.8–1.2	0.7–1.1

Conversion processes: Injection moulding, extrusion, blow moulding.

Uses: PA 6, PA 6 6 and PA 6 10: Components subject to wear like gears, sprockets, axles, ball bearing races, bushes, couplings, shoe heels, lock parts, load bearing articles like nuts and bolts, handles, levers, ventilators and machine components, lamps and battery housing, bobbins, combs etc. (PA 6 10 has better dimensional stability). Automobile components like oil tubs, fans, pedals, hub caps, sliding roof frame, airbag housing and technical components under the hood.

PA 11 and PA 12: Gears, bearings gaskets, housings and other components coming in contact with water.

Special note: Polyamides are semi-crystalline and as such opaque but the aromatic polyamides are transparent.

The number given after the abbreviation PA denotes the number of carbon atoms in the monomer (PA 6, PA 11, PA 12). The double number indicates a monomer with two molecules and stands for the number of carbon atoms in each molecule (PA 6 6, PA 6 10).

Polyacetal (POM = Polyoxylmethylene), is an engineering thermoplastics from the semi-crystalline family. Its distinguishing properties as listed below make it an ideal material for technical applications.

- High hardness and stiffness.
- High toughness, also up to −40°C
- High heat resistance
- Low water absorption, good dimensional stability
- Good electrical properties
- Resistant to petroleum, oils, fats, alcohols, weak acids and alkalis, detergents.

- Excellent surface hardness and wear resistance
- Very low friction, self-lubricating behaviour
- Polyacetal is whitish in natural state. It burns with a pungent bluish flame smelling of formaldehyde.

Special attribute: Polyacetal is suitable for long lasting integral hinges.

Specific gravity	1.42
Moulding temperature	180–230°C
Service temperature without load	
Short term	110–140°C
Long term	90–110°C
Mould shrinkage	1.9–2.3%

Conversion processes: Injection moulding, extrusion, blow moulding.

Uses: Components for cars, office and household machines, gears, bearings, screws, parts for pumps, textile machinery, radio, television, aerosol components, springs, engineering components with snap fittings, parts of conveyor belts, etc.

Polyetheretherketone (PEEK), is a high performance semi-crystalline thermoplastics. Its high heat resistance and mechanical properties can be further enhanced by reinforcement with glass or mineral fibres. The distinguishing properties of PEEK are:

- Very high tensile strength
- Very good fatigue resistance
- Very high heat resistance
- Very good chemical resistance, also at elevated temperatures (soluble only in sulphuric acid).
- Excellent electrical properties
- Good wear resistance
- Low friction
- Low water absorption
- Resistant to hydrolysis
- Self extinguishing

Specific gravity	1.32
Moulding temperature	375–450°C
Service temperature without load	
Short term	315°C
Long term	260°C
Mould shrinkage	0.7–1.2%

Conversion processes: Injection moulding, extrusion.

Uses: Electrical and electronic components, connectors, relays, printed circuits, pumps for aggressive fluids as well as for foodstuff, replacement of metal components in automobiles, aeroplanes and rockets as well as in military hardware.

Special note: Polyetherketone (PEK) polyaryletherketone (PAEK) have similar properties.

Liquid Crystal Polymers (LCP), are a unique class of thermoplastics possessing properties of semi-crystalline and amorphous plastics and much more. Based on polyester, their molecules are stiff, rod-like structures, organised in large, parallel arrays in the melted as well as solid state. Their molecules lend strength like reinforcing fibres and hence LCPs are also called self reinforcing plastics.

- They are resistant to most solvents, even at elevated temperatures.
- They have a low melt viscosity; their L/T ratio of flow may exceed 350.
- Their cooling time is very short.
- They have the least shrinkage and warpage of all thermoplastics.
- They have very high tensile strength, stiffness and hardness.
- They have good creep resistance.
- They retain their physical properties even after repeated heating and cooling.
- Their natural colour is beige.
- They are self-extinguishing.

Special attribute: Some filled grades exhibit negative shrinkage along flow.

Specific gravity	1.4
Moulding temperature	390–410°C
Service temperature without load	
Short term	260–300°C
Long term	220–240°C
Mould shrinkage:	0.3–0.6%

Conversion processes: Injection moulding.

Uses: Components for electrical field such as plugs, connectors, socket switch components, microwave appliances, coil bobbins.

Components for automotive sector such as parts for fuel, ventilation and air conditioning.

Components for medical field such as parts for hearing aids, metal substitution in surgical and dental instruments, sterilisation containers, dosing appliances etc.

General: Gears, computer parts, sensors, parts of printing machines, axle of photostat machine, pump housing, mobile phone parts, fibre optics, CD players, etc.

CLASSIFICATION OF THERMOPLASTICS

On the basis of their properties and application, thermoplastics are classified in three categories:

Commodity Plastics

The group of thermoplastics, which can be safely used up to 90°C, is known as commodity plastics. It comprises polyvinyl chloride, polystyrene and its derivatives and the polyolefines. ABS and PP, however can also be used with boiling water.

Commodity plastics find use in packaging, household articles and home appliances.

Volume wise, commodity plastics form the largest part of processed thermoplastics. Extrusion, injection moulding and blow moulding are the major processes for their conversion. The maximum processing temperature for commodity plastics does not exceed 260°C. Only a few of the commodity plastics, e.g. HIP, SAN and ABS are hygroscopic and call for pre-drying.

Engineering Plastics

Thermoplastics, with higher temperature resistance than the commodity plastics and with better mechanical properties, for example, polycarbonate, polyethleneteraphthalate as well as polybutyleneteraphthalate and polyacetal fall in this group, so do polysulphone and polyamides. Engineering plastics may be subjected to temperatures as high as 140°C. PSU and PA 4.6 can go up to 160°C.

Engineering thermoplastics have much better mechanical properties than the commodity plastics and are used for manufacturing components for technical application. They are mainly processed by injection moulding. Their processing temperatures range between 250–350°C. Most of them need pre-drying before moulding. The moulds too are maintained at higher temperatures.

High Performance Thermoplastics

This special class of plastics is superior to the other two groups not only in very high temperature resistance but also in electrical, mechanical and chemical properties.

These include polyetherimide, polyether sulphone, polyphenylene sulphide, liquid crystal polymers, polyether etheramide and polytetrafluoroethylene. They can withstand temperatures up to 260°C.

High performance thermoplastics are processed exclusively by injection moulding. The processing temperatures exceed 400°C in certain cases. The injection pressures too range from 1000 to 2000 bar. The moulding machines for their processing must have special plasticising units to withstand high temperatures and wear. The injection moulds for high performance thermoplastics have to be heated to 120°C and above.

THE FORM OF THE RAW MATERIAL

Plastics are generally compounded by the raw material manufacturer and supplied ready for processing into desired products. The most common form of the thermoplastics raw materials is granules or beads, pigmented or in their natural colour. Their form facilitates easy transport and feeding into the processing machinery.

Polyvinyl chloride, a thermoplastics, is also supplied as powder. It enables the processor to make a formulation answering his particular requirement by mixing various additives in proportions yielding the desired mechanical and thermal properties. PVC is a very versatile polymer which can be compounded to give numberless variations in rigidity from brittle hard to soft rubbery state.

Thermosetting plastics are available as granules, powder and dough for various methods of processing.

PREPARATION OF THERMOPLASTICS FOR MOULDING
Pre-drying

Some of the thermoplastics are hygroscopic. It is imperative that they be dried before moulding. The moisture leads to defective mouldings and rejection. All plastics, however, cannot be pre-dried at the same temperature. Table 1.3 lists the recommended drying temperatures and the duration for various hygroscopic thermoplastics.

The reground materials should be dried separately and stored in airtight containers.

The most effective way of drying the polymer granulates is by circulation of pre-dried hot air.

Table 1.3: Pre-drying of Thermoplastics			
Acronym	*Polymer*	*Pre-drying Temp. °C*	*Duration hours*
SB	High impact Polystyrene	70	2–3
SAN	Styrene Acrylonitrile	70–80	2–3
ABS	Acrylonitrile Butadien Styrene	70–80	2–4
CA	Cellulose acetate	70	3
PMMA	Polymethyl Methacrylate	70	4
PC	Polycarbonate	120	4
PPO	Polyphenylene Oxide	100	2
PSU	Polysulfone	150	6
PET	Polyethylene Terephthlate	130	2–4
PA 6	Polyamide 6	80	2–4
PA 6 6	Polyamide 6 6	80	2–4
LCP	Liquid Crystal Polymer	150	1–2
POM	Polyacetal	110	2–3
PEEK	Polyetheretherketone	150–200	5
PC-ABS-Blend	Polycarbonate-ABS-Blend	90–110	2–4
EVA	Ethylene Vinyl Acetate	70	1–2

Colouring of Thermoplastics

Thermoplastics are rarely used in their natural colours. The natural colour of semi-crystalline thermoplastics is white or pale yellow. The amorphous plastics are transparent like glass with a yellowish or bluish tinge. A positive attribute of plastics is their ability to take on almost any colour.

Colorants for plastics are of two types. Pigments are generally inorganic compounds, mostly compounds of metals like titanium, iron, chromium, cobalt, cadmium, etc. They do not dissolve in plastics melt but lend it the corresponding colour and make it opaque. They can usually withstand high temperatures during processing and application. Titanium dioxide is the most commonly used pigment for making plastics white or opaque. Some of the pigments are toxic in nature and may not be used for products coming in contact with edibles or the human body.

Some commonly used pigments are:

White   —   Titanium dioxide, zinc sulphide
Yellow   —   Lead chromate
Red   —   Iron oxide, molybdenum red
Green   —   Chromic oxide, cobalt green
Brown   —   Iron oxide
Black   —   Carbon black

Golden, metallic, pearlescent and fluorescent colours constitute the specialities.

Dyes are organic substances and get dissolved in plastics. They are transparent and when used with natural amorphous thermoplastics, lend it their colour without making it opaque. Although more brilliant in colour, dyes are more heat and light sensitive and some may also bleach out gradually in use. Because of these limitations, dyes are not used as extensively as the pigments.

Thermoplastics are available in their natural colours. They are also supplied by the producers and compounders in a large range of standard colours at an additional cost. Even special colours and hues can be ordered if the quantity is large enough.

Inspite of the higher price, the ready for use, coloured plastics assure uniformity, homogeneity and repeatability from batch to batch because the suppliers have at their disposal, the expensive machinery for compounding, controlling and testing and the know-how and experience. Colouring in house is not more economical in most cases but it may have to be resorted to for out of-the-way colours and shades, especially when the quantities required are not very large.

Choice of colorants: Colorants effect not only the appearance but also the morphology of the plastics. Depending upon their own chemical structure, they influence the melting and rheological behaviour, the crystallinity and the shrinkage. The colorants must fulfil following criteria:

• Be compatible with the polymer
• Withstand the processing temperature
• Should not plate out leaving a layer of colour on mould surface.
• Withstand working temperatures for prolonged periods
• Remain unchanged under operating environmental conditions, such as chemicals, gases and fumes, UV rays, moisture, etc.
• Should not migrate or bleach out
• Should not be effected by light
• Should not alter the mechanical properties of the polymer.

Colouring in Own Shop

Because of the following reasons, the moulders may choose to do the colouring of plastics themselves:

• The desired colour is non-standard and the quantity required is not large.
• The product has to be moulded out of the same material but in a number of colours.
• The material in natural colour is cheaper when purchased in bulk.
• The inventory of the raw material in only a few colours is simpler and more economical.
• Ease of storage. Space saving

Various options for colouring the raw material in house are:

Dry Colouring

It is carried out with pigments in powdered form. A precisely weighed amount of the colour is added to a predetermined quantity of plastics granules along with a wetting agent in liquid form. The mixing takes place in a tumbler which consists of one or two hollow drums with tightly fitting closures. The drums are revolved around in a vertical plane at slow speed by means of an electrical motor. The design of the equipment facilitates constant mixing of granules and the powdered colour.

Advantage: The apparatus is simple in design and low in cost. The mixing is fairly uniform. It is quite useful for small quantities which could be used immediately. The quality of colouring in the product is satisfactory.

Disadvantages: The process must be carried out in isolated rooms, as the powder may contaminate the surroundings. The apparatus must be kept meticulously clean if it is to be used for various colours. The pigment may stick to the screw and the barrel of the plasticising unit and may contaminate the subsequent charges in another colour.

The colour may separate and settle down during long storage and transport over long distances. The colouring of the mouldings may not be uniform in case of separation of pigment. Moreover, the colour may not disperse equally to thin sections of the moulding.

The mixture cannot be transported to the machine automatically.

Wet Colouring

Instead of powdered pigment, the colouring substance is in the form of paste. Mixing with the granulate may be done in a tumbler for small quantities and the mixture poured in the hopper or the pigment may be added to the granulate right on the moulding machine. It is effected by a dosing gadget coupled with the hopper. The predetermined quantity of the paste is pumped in with each cycle.

Advantages: Wet pigmentation is an economical method. The pigment does not pollute the surroundings. The pigmentation is more uniform. The quality of mouldings is very good, especially when dosing is carried out accurately on the machine.

Disadvantages: A dosing gadget is necessary for every machine for regular production. The pigment sticks to all parts of the plasticising unit. The dosing gadget as well as the injection unit must be thoroughly cleaned before colour changes. By higher concentrations of the pigment, separation may take place which can bring colour fluctuations in production. The pigment may not disperse equally to thin sections of the moulding.

Colouring with Master Batches

Master batches are colour concentrates, consisting of pigments embedded in thermoplastics. They are prepared by mixing large quantities of the pigment with the plastics and regranulating it. The percentage of the pigment may be as high as 50%. The carrier plastics is compatible with the plastics to be coloured.

Granules of the master batch can be mixed with material granulate in required proportion either in a tumbler or the batch can be metered to the raw material on the machine with a volumetric dosing gadget.

Advantages: The process is very clean at all stages. It does not pollute the environment, the mixer, the machine and the mould. Cleaning and change of the colour are very simple. Dispersion of the colour is uniform. The dosing is simple and accurate. Through higher dosage, greater colour intensity is possible.

Disadvantages: Master batches are more expensive than the dry colours or pastes. They cannot be set in universally as the carrier plastics in the master batch must match with the polymer to be coloured. The uniformity of distribution of colour depends upon the efficacy of the screw. The colour dispersion in thin sections of the moulding may not be adequate.

It must be remarked in conclusion that the geometry of the screw also plays a crucial role in satisfactory mixing and dispersion of the pigments. General purpose screws under 20D length do not deliver adequate results. Either screws with additional mixing head or those with a length of 25D are recommended for colouring on the machine.

Dip Dyeing

Although a post operation, dip dyeing may replace pre-colouring in many cases. Recently developed by Bayer Material sciences of Germany and christened "Aura", the patented process consists of dipping the moulded or extruded article in an aqueous solution containing the dyes along with specific surficants. The solution may be prepared by dissolving the solvents and the dyes in deionised water at 50–60°C and then heating it up to 96°C.

The dyes get diffused in the product in a very short period which may range from a few seconds to some minutes depending upon the depth of penetration intended. Variation in wall thickness does not effect the appearance of the treated product. Special features like metal flakes or sparkle present in the article, are enhanced by dip dyeing. The last operation in dip dyeing comprises rinsing of the dyed article in clear water and drying with hot air.

The process may be applied to most amorphous polymers and to some semi-crystalline ones too.

The economical advantages of the dip dyeing technology come to the fore when a product has to be manufactured in a number of colours in small batches. The conventional moulding process would involve change of material by purging for every new colour, resulting in wastage of material as well as production time. The dip dyeing method obviates storage of plastics raw material in numerous colours and thus reduces inventory costs considerably.

The process can also be employed to impregnate moulded/extruded articles with UV-stabilisers, antistatic agents, etc., also in conjunction with dyeing.

RECYCLING

Theoretically, thermoplastics can be remelted and moulded any number of times. It must, however, be emphasised that some degeneration does take place with each round of melting, exercising adverse influence on certain properties. Reground material can be added to the virgin one in certain proportion without discernable difference in quality of the mouldings. Twenty percent is considered to be the safe limit. The size of the reground particles should not differ much from that of the virgin granules and the mixing should be quite thorough so that the resultant melt for each shot is uniform.

Mixing of recycled material is not recommended for applications where any decline in properties may result in failure of the product or risk of health, such as with articles for medical application, etc.

Thermosetting plastics cannot be recycled for moulding. They can, however, be used as fillers with materials in other fields.

ACRONYMS

Plastics are generally referred to under their acronyms, a list of which is given below.

Acronyms for Some Common Polymers

ABS	Acrylonitrile Butadiene Styrene
ASA	Acrylester Styrene Acrylonitrile
CA	Cellulose Acetate
CAB	Cellulose Acetate Butyrate
CAP	Cellulose Acetate Propionate
CN	Cellulose Nitrate
EVA	Ethylene Vinyl Acetate
FEP	Fluorinated Ethylene Propylene
HDPE	High Density Polyethylene
HDHMWPE	High Density High Molecular Weight Polyethylene
HIPS	High Impact Polystyrene
LCP	Liquid Crystal Polymer
LDPE	Low Density Polyethylene
LLDPE	Linear Low Density Polyethylene
MF	Melamine Formaldehyde
PA	Polyamide
PAI	Polyamide Imide
PAN	Polyacryle Nitrile
PBTP	Polybutylene Terephthalate
PC	Polycarbonate
PEEK	Polyether Ether Ketone
PEI	Polyetherimide
PES	Polyether Sulphone
PET/PETP	Polyethylene Terephthalate
PF	Phenol Formaldehyde
PI	Polyimide
PMMA	Polymethyl Methacrylate
POM	Polyoxymethylene

PP	Polypropylene
PPO	Polyphenylene Oxide
PPS	Polyphenylene Sulphide
PS	Polystyrene
PSU	Polysulphone
PTFE	Poly Tetra Fluoro Ethylene
PU/PUR	Polyurethane
PVC	Polyvinyl Chloride
PVDC	Polyvinylidene Chloride
PVDF	Polyvinylidene Fluoride
SAN	Styrene Acrylonitrile
SB	Styrene Butadiene (HIPS)
TEEE	Thermoplastic Ether Ester Elastomer
TPE	Thermoplastic Elastomer
TPU	Thermoplastic Polyurethane
UHMWHDPE	Ultra High Molecular Weight High Density Polyethylene
UF	Urea Formaldehyde

2

Injection Moulding of Thermoplastics, the Equipment

- History
- The injection moulding process
- The moulding equipment

HISTORY OF INJECTION MOULDING

Shellac, a natural tropical resin, had been in use in the western countries since the middle ages but it was around 1850 that a method was developed for mass production of articles out of it. It involved liquefying of the solid resin by application of heat and forcing or "injecting" the melt into a confined cavity having the shape of the intended product. The "cavity" was usually contained in two metal blocks, which could be held together during injection and solidifying of the molten resin and then taken apart to extract the solidified product. The arrangement was called a mould. The existing process of die casting served as a model.

The first injection moulding device for the natural thermoplastics was a modified die casting machine. The range of products was limited to some household utility articles and decorative items. It was the invention of celluloid in 1872 which spurred development of the injection moulding process. The first patent for an injection moulding machine for thermoplastics was registered by the Hyatt brothers in the United States in 1872. It was a vertical, manually operated machine, not much different from that still being used in India for simple, small mouldings in the cottage industry (Fig. 2.1). A fact, which helped penetration of the new material in many areas, was the availability of expertise in mould making in the field of glass blowing. Celluloid became a synonym for thermoplastics and many utility articles appeared in the new material in attractive colours and affordable prices. Plastics toys became a rage and ruled the market till the first world war. More than the material, the process of its moulding had brought a revolution in the whole concept of mass production.

The next notable development took place in 1921. Hermann Buchholz of Germany introduced an improved version of the manually operated injection mouldings machine. Eckert and Ziegler, German machine manufacturers, replaced the manual injection by pneumatic force and introduced their invention in 1925. Essentially, it was a vertical machine like its predecessor and the

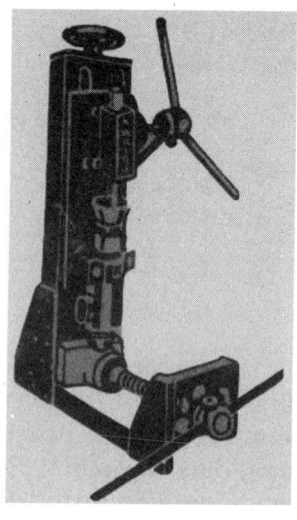

Fig. 2.1

mould closing was still manual. Eckert and Ziegler became leading moulding machine manufacturers though many more followed soon. The truly automatic era dawned when the plasticising unit was put in horizontal position in the thirties and the important parameters like pressure and durations of injection, cooling, etc. could be pre-set. The plasticising unit consisted of a hollow steel barrel, equipped with band heaters for melting the granules, and a piston or a "plunger" fitting closely in the barrel for injection, powered pneumatically, electrically or hydraulically. Plasticising was improved by introducing a torpedo in the barrel in the path of the melt, so that it received heat not only from outer band heaters but also from inside, though indirectly (Fig. 2.2). Later, the torpedoes too were equipped with internal heaters.

A revolutionary development was the invention of reciprocating single screw plasticising and injection. The screw fulfilled functions of conveyor, plasticiser, mixer and of the injection plunger. The first screw preplasticising injection moulding machine (Fig. 2.3) was introduced in 1959 by Ankerwerke of Germany and the principle has given a tremendous boost to plastics.

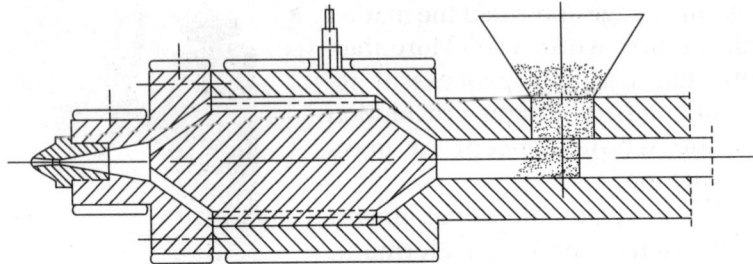

Fig. 2.2

Thermosets had been processed, since their advent in the first decade of the twentieth century, on vertical presses by semi-automatic processes of compression and transfer moulding. It was in 1963 that a screw pre-plasticising automatic machine was introduced for moulding of thermosetting plastics.

The in-line screw pre-plasticising and injection system has become a norm and has experienced numerous improvements since its introduction. The in-depth studies and research in the behaviour of plastics melts have contributed to refinement of the functions of the screw pre-plasticising and injection. Computerisation has brought very precise setting, repeatability

Fig. 2.3

and control of the parameters. The closing systems, too, have been refined and extended to electro-hydraulic, servo-hydraulic, fully electric and combinations. Tiebarless and two platen versions of mould closing systems and the closed loop injection have extended the scope, productivity and accuracy of the injection moulding process tremendously.

The injection moulding process has not remained confined to thermoplastics, as would be seen in the ensuing chapters.

THE INJECTION MOULDING PROCESS

Injection moulding is rightly regarded as the most versatile process for conversion of thermoplastics. It is perhaps the only process which manufactures even the most intricate articles, right from the raw material to the final form, in one shot, obviating after-operations in most cases.

Basically, the injection moulding process consists of two distinct steps.

a. Plasticising or liquefying of the solid plastics
b. Transferring the melt to a mould giving it the shape of the intended product and stabilising it in that form.

In practical terms, the process comprises the following operations.

- Plasticising the thermoplastics raw material, i.e. converting it from solid to fluid through heat energy.
- Injecting the plasticised material in a confined cavity of the shape of the article to be formed.
- Extracting the heat from the injected melt to resolidify the plastics in the given shape.
- Opening the cavity and extracting the solidified moulded article.

The process is carried out by an injection moulding machine, which performs these functions and converts the injected plastics into the intended article with the help of a mould.

THE MOULDING EQUIPMENT (FIG. 2.4)

In order to be able to perform various steps involved in the injection moulding process, the equipment should have the means to:

- Plasticise the polymer
- Hold and close a mould

- Inject the plasticised plastics with pressure in the closed mould
- Cool the injected material
- Open the mould
- Eject the moulding

The equipment, viz. the injection moulding machine, consists of several units which perform the above named tasks. These are:

- The plasticising unit
- The clamping unit
- The ejection device
- The cooling battery
- The machine body
- The control panel

Figure 2.4 shows above units in their usual position.

The Plasticising Unit (Fig. 2.5)

It transports the material from the source (feeding point) towards the delivery (nozzle) end.

It plasticisers (melts) the solid polymer and homogenises it while conveying it to the nozzle end.

It injects the melt into the mould.

It doses or recharges the material for the next injection.

The plasticing unit consists of the following parts:

The Barrel

The plasticising unit consists of a long, hollow metal cylinder called barrel, which contains a screw closely fitting in its bore

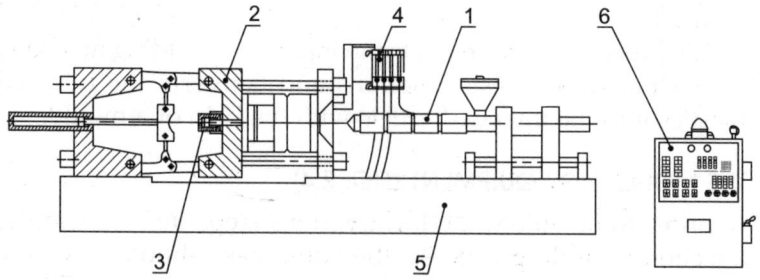

Fig. 2.4

(Fig. 2.5). The screw can rotate and reciprocate. The cylinder, which is generally placed horizontally, is fixed on one end to a motor which powers the screw. The barrel has a vertical opening at this end of its periphery, where a hopper for the raw material is mounted at right angles to its axis. It resembles a conical funnel, circular or square in cross section, and is generally fabricated out of stainless steel sheet. It has a closely fitting lid, which is sometimes foreseen with electrical heaters on the inside to prevent moisture deposition on cold plastics raw material. The hopper is receptacle for the granules of the thermoplastics to be moulded. It has a glass covered side window to display the level of the filling. It has also provision for shutting off the aperture to the barrel and for emptying of the contents.

The barrel is equipped with a number of electrical band heaters controlled by means of thermocouples, which are placed in holes drilled in the wall of the barrel to a safe depth but not entering the material bore. However, the section of the barrel around the hopper (the feeding section) is provided with water cooling to prevent the material granules from getting heated, melting superficially, sticking together (bridging) and blocking the passage from hopper to the screw.

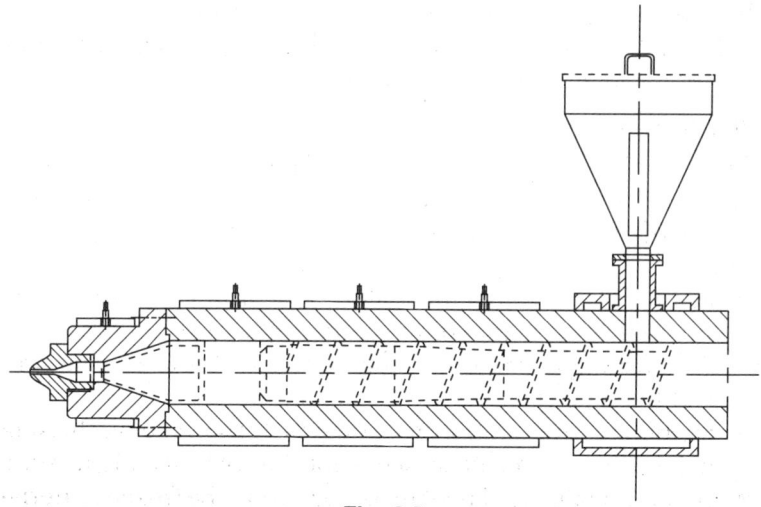

Fig. 2.5

The Cylinder Head

The end of the barrel opposite to the feeding zone is closed by a flange like component called the cylinder head, centred and fixed onto the barrel with screws. Its function is to reduce the bore of the barrel gradually to that of the nozzle placed on its exit end. The nozzle is generally screwed into the cylinder head. Like the barrel, the head and the nozzle are also equipped with band heaters and thermocouples. The cylinder head for machines processing PVC and thermosets is designed to be readily detachable.

The barrel, the cylinder head, the nozzle and the hopper form the outer components of the plasticising unit. The barrel and its head are made of nitriding steels for processing of common polymers without abrasive fillers. Stainless steel is preferred for aggressive polymers like PVC and POM and a bi-metal hard sleeve inside the barrel is used for materials containing abrasive fillers.

The nozzle forms the end of the barrel and represents the link between the plasticising unit and the injection mould. Its inner bore tapers down to a small orifice in the range of 3–6 mm., through which the molten plastics is injected into the mould. The plasticising unit assembly is generally placed horizontally. It can be moved linearly to and fro.

Special purpose machines may have the plasticising unit in positions other than horizontal also.

The Nozzle

The machine nozzle forms the exit end of a plasticising unit and is fitted onto the cylinder head. It is the link between the source of the melt and the receiver, viz the mould. Its functions are:

- Reducing the size of the melt channel from the bore of the cylinder head to that of the entry in the mould.
- Establishing a leak proof link between the plasticising unit and the mould.
- Enabling a smooth transfer of the melt from the plasticising unit to the mould with least possible loss of heat and pressure.
- Separating the fluid melt in its body from the frozen material of the sprue in the mould.

- Preventing drooling of the melt from the plasticising cylinder, when the unit is not resting on the mould.

There are a few designs of the nozzle for general and specific requirements. The most common types are:

The open nozzle (Fig. 2.6). An open nozzle is the standard outfit for an injection moulding machine and is used with most plastics. It is centred and screwed in the barrel head. As the figure shows, its internal bore, starting from the internal diameter of the cylinder head, tapers down to a smaller hole to match with the sprue. The small hole runs straight for 3–5 mm. and then funnels out. This measure forms a distinct tear-off point where the cold material in the tip of the nozzle resting against the sprue bush of the cooled mould would tear off from the hot melt on the right and go out with the frozen sprue.

The open nozzle, besides being the simplest and the cheapest, provides a smooth, interruption-free passage to the melt, causing very little pressure loss. It does not have any dead corners where the melt could stagnate and degenerate. It is suitable for almost all materials, particularly for the heat sensitive ones. Polymers with very low viscosity, such as the polyamides, pose a problem of drooling with the open nozzle. Special nozzles with shut-off mechanisms have been developed for use in such cases. These are:

The spring operated shut-off nozzle: The nozzle is closed by a spring loaded needle or pin placed in the central path of the melt. The closing end of the pin is tapered so that at the time of injection, the melt pressure pushes it back and sets the passage

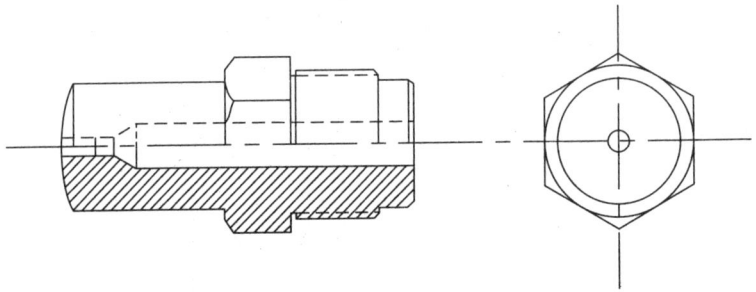

Fig. 2.6

free for the melt. As the needle has to be placed on the central axis in line with the bore of the nozzle, the melt from the barrel must be diverted around it. The flow is no more streamlined as in an open nozzle.

The version shown in Fig. 2.7 has an externally placed helical spring, which exerts pressure on the needle through a crossbar. A shortcoming of the nozzle is that the material channels cannot be distributed uniformly due to the space occupied by the cross bar. However, the spring can be replaced without taking the nozzle apart.

Another version, depicted in Fig. 2.8, differs from the foregoing design in the location of the spring. The working principle is identical in both versions. The spring (a number of disc springs) is located inside and the melt goes around the spring housing through a number of evenly distributed channels. All material streams meet again near the central bore of the nozzle blocked by the conical tip of the needle. The drawback of the design is that the replacement of springs requires dismantling of the nozzle.

As the spring operated needle valves depend upon the injection/holding pressure to open the melt passage, the nozzle can remain partially closed if the pressure is low. A part of the pressure can also be lost in overcoming the force of the spring. Charging or refilling for the next shot is possible only with the

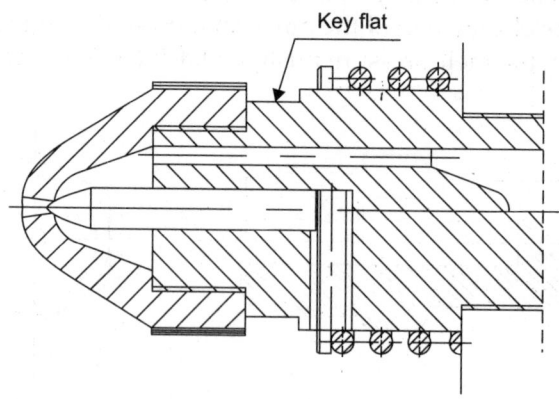

Key flat

Fig. 2.7

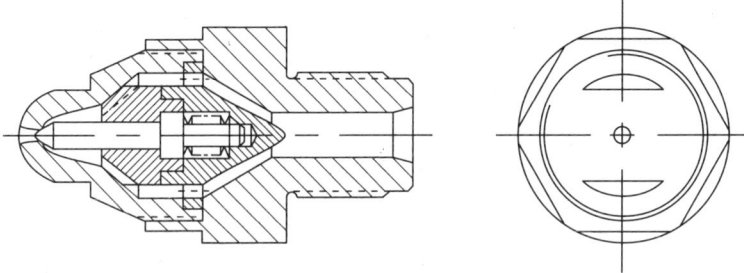

Fig. 2.8

nozzle pressing on the sprue bush as otherwise the back pressure may push back the closing spring leading to drooling. Springs, working constantly at high temperature, may lose a part of their tension after some time.

Spring loaded nozzles separate the hot melt from the colder material in the front bore which goes out with the sprue. Another advantage offered by these nozzles is the possibility of charging as well as purging without the support of the mould.

Hydraulically operated shut-off nozzle (**Fig. 2.9**): The blocking of the melt path is effected by a cylindrical bolt, placed across the main bore. The bolt has a hole corresponding to the material passage, which can be brought in line by turning the bolt in "open" position, by means of a small hydraulic cylinder, so that the melt can pass through unhindered. A modified version slides the bolt to align its hole with the melt bore, instead of turning as described before.

This design offers the best solution as there are no divisions, no diversions and no narrow channels which could overheat the material. The opening and closing is regulated by the machine controls.

The slider shut-off nozzle: The main part of this shut-off nozzle is a tip which can slide linearly in the nozzle body which, in turn, is fixed onto the cylinder head. Figure 2.10 demonstrates its mode of functioning. When the injection unit is not in contact with the mould, the sliding tip of the nozzle is kept in the forward position by the melt pressure so that the angular passages of the melt to the central outlet path in the tip are

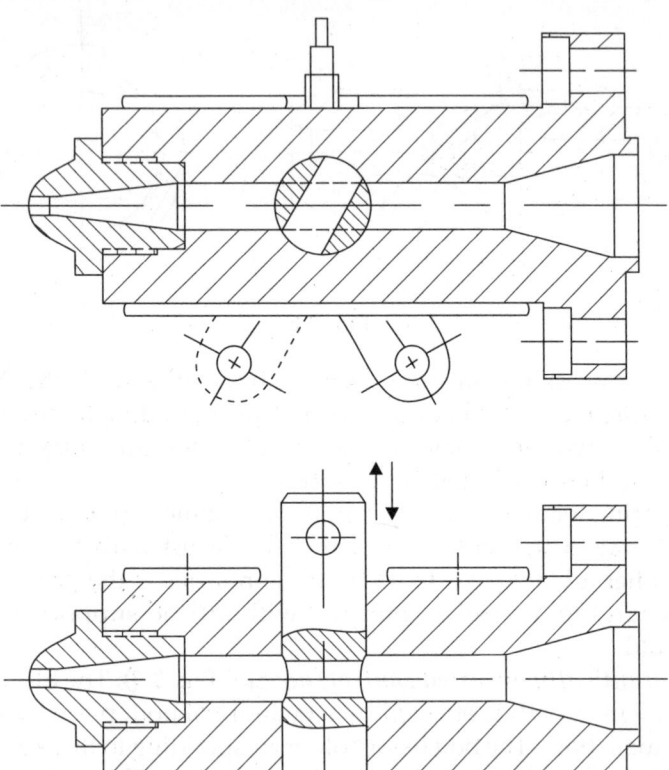

Fig. 2.9

blocked. When pressed against the sprue bush of the mould prior to injection, the tip is pushed to the rear position and the angular channels get connected to the central bore of the nozzle body. Now the melt can be conveyed to the mould. When the cylinder is pulled back after the holding phase, the material pressure pushes the tip forward and shuts off the angular channels.

The nozzle is simple in design and does not require external aids such as springs or hydraulic cylinders for operation but suffers from the drawbacks caused by diversion. The melt left

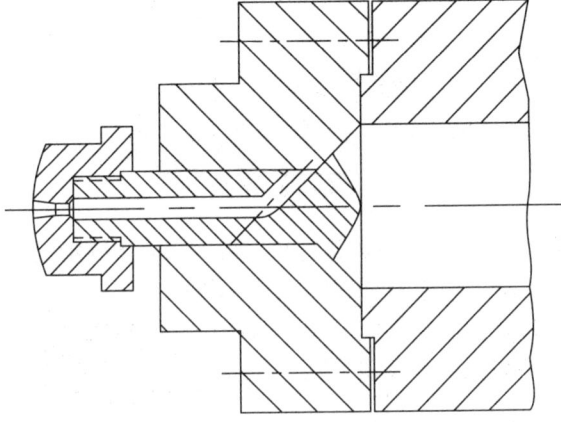

Fig. 2.10

in the bore of the sliding tip after the shut off may cool and lead to blockage or to the defects in the subsequent moulding caused by cold material. Purging is possible only with some additional device.

It may be mentioned that apart from the designs described above, there are numerous variations of the shut-off nozzles, developed by different machine manufacturers.

Static Mixers: Although the screw also effects mixing, static mixers may be incorporated in the machine nozzle to enhance the degree of mixing, especially of the pigments added to the granulate in form of liquid, powder or master batch. Static mixers are devices to divide the melt into a number of smaller streams which join again at the end. The pineapple mixer (Fig. 2.11) is the simplest form of a static mixer. The other designs employ zigzag plates or barriers to the same end (Fig. 2.12). It proves beneficial with sophisticated products where the surface finish is one of the prime quality attributes. The more uniform melt temperature narrows down the dimensional fluctuation and thus helps reduce rejection rate. Static mixers are also needed for moulding of thermoplastic blends.

Depending upon the design of the mixing channels, there is some loss in the injection pressure. The temperature of the melt,

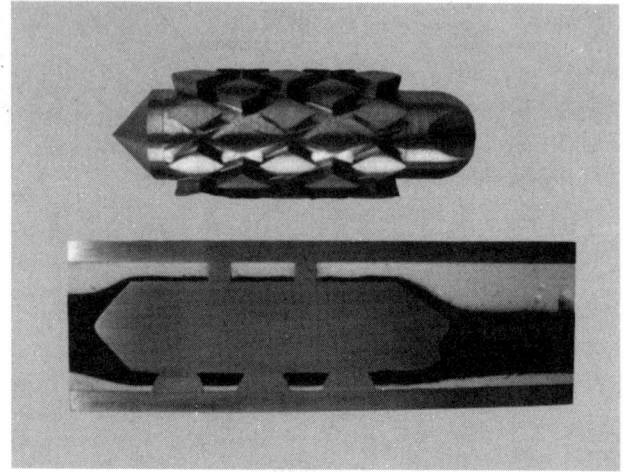

Fig. 2.11

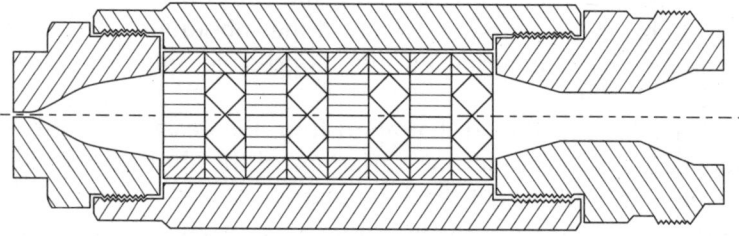

Fig. 2.12

however, experiences a rise. Generally, the injection pressure does not require compensation as the melt with more even temperature distribution flows more easily.

Mixers, however, come in the way of the suck-back.

The Screw

The primary function of the plasticising unit, suggested by its name, is executed by the screw but apart from plasticising, the screw performs a number of tasks as listed below:

- Transporting the raw material from the source (hopper) towards the delivery point (barrel head).
- Compacting the material and expelling air

- Plasticising the raw material through friction
- Mixing the material
- Homogenising and compressing the molten material
- Injecting the melt in the mould
- Compensating for the shrinkage of the melt in the mould

In order to be able to perform these functions, the screw is given a special configuration. It is divided in three distinct zones or sections along its length. These are feed zone, compression or the transition zone, and the metering zone, placed one after the other in the named order (Fig. 2.13). They differ from each other in their geometry and their functions.

The Feed Zone

The feed zone of the screw (A) is situated right under the hopper from which it gets the material and transports it further to the next zone when it rotates. It extends over half the length of the screw. The threads of the screw have their maximum depth in the feed zone. It is usually 0,15 times the diameter for the general purpose screws. The minimum depth of the feed zone, however, is never less than 3 mm. so as to be able to accommodate plastics granules. This is why, screws of a diameter less than 12 mm. are not feasible. A core diameter of less than 6 mm. will not be able to withstand the torque acting upon it during operation.

The Compression Zone

The compression zone (B) starts where the feed zone ends. The depth of the thread starts decreasing gradually in this zone. This leads to compaction of the granules and results in expulsion of the included air towards the hopper. As the sectional area decreases gradually, the material is compressed more and more. Consequently, the friction between the

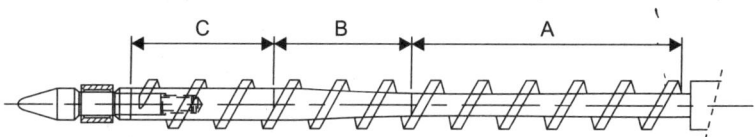

Fig. 2.13

granules and the barrel and also among the material particles increases and the plastics starts melting. A part of the heat comes from the electrical heaters fitted around the barrel. A molten film deposits on the bore of the barrel but the advancing flank of the rotating screw keeps on scraping and pushing it inwards in a looping motion. This action leads to mixing and uniform heating of all material particles.

The length of the compression zone is about one fourth of the length of the screw. The ratio of the thread depth at the start and at the end of the compression zone is termed as the compression ratio (actually the compression ratio relates to the reduction in volume but as the pitch of the thread remains constant, the ratio of depth corresponds to that of the volumes in the beginning and at the end of compression). This is the factor having the maximum influence on friction and heat generation. The compression ratio of the general purpose screws ranges from 2.5:1 to 3.5:1. For heat sensitive materials like PVC, it does not exceed 2:1. As a rule, a higher compression ratio increases the output. For polyolefins, which have a wide processing temperature range, it may exceed 4:1. It is lower for more viscous materials such as polycarbonate.

The Metering Zone

The metering zone (C) forms the last section of the screw and begins where the compression zone ends. It has a uniform depth all along, namely that at the end of the preceding zone. It is usually about one fourth of the total screw in length. Its main function is to stabilise and homogenise the melt and bring it to the final temperature for injection. As the depth of the screw is minimum in this section, there is maximum friction between the melt and the cylinder wall as well as between various layers of the melt itself. The sticking and scraping during the rotation leads to constant mixing and homogeneity of the temperature.

A non-return valve fitted at the end of the metering zone forms the last part of the screw. As described in the next section, the melt being forwarded over the metering zone by the rotating screw can pass through the valve towards the barrel head but cannot flow backwards when the screw acts as an injection plunger. As the melt flows forward during plastication, the screw recedes back, making room for the melt.

The other function of the screw is injection of the melt. It acts as a linearly moving plunger for performing this task. It is pushed forward to force the molten material out of the barrel. The non-return valve prevents leakage of the melt in reverse direction. The screw does not rotate during the injection stroke.

The dimensions of various features of the screw are linked with its diameter. The pitch of the screw is generally maintained uniform as 1xD throughout its length. The helix angle works out to about 17.8, which has been found to be most efficient for transporting the material. The landwidth of the thread flank is taken as one tenth of the screw diameter. Its front flank, pushing the material, is kept vertical to the screw axis and joins the root diameter with a fillet radius whereas the back of the flank is tapered at an angle of 20–30 degrees to increase its strength. The ratio of the length to the diameter of the screw may vary from 18 to 24. The larger the L/D ratio, the more the shear heat generated by friction and better the mixing.

Too long a screw is difficult to support and maintain straight. It may start touching the barrel due to the overhang. The length influences the plasticising distance and hence the degree of homogeneity. Too short a distance may effect the homogeneity adversely. The length of various sections is expressed as a function of the screw length.

Irrespective of the length, the screw and barrel must have a very close fit. The maximum clearance aimed at is about 0.2 mm.

Screws with shallow depths are better for homogeneity but lower in output.

The screws are hardened by nitriding. The hardened layer measures about 0.2 mm. The material used for fabrication of screws is a nitriding steel. Coatings of special hard alloys are deposited on the screw surface for processing of filled, abrasive compounds. Wear and tear in the plasticising unit occurs due to removal of metal particles from the surface by the abrasive action of fillers in the plastics. Adhesive wear is caused by the strong holding power exercised by plastics like polycarbonate, polysulfones, etc., especially at lower temperatures. The thin nitrided layer from the surface of the screw gets removed gradually and the clearance thus generated enhances backward leakage. Subsequently, the efficiency decreases with increasing wear.

Another serious cause of the wear of the plasticising unit is decomposition of certain heat sensitive plastics. Some volatile additives, too, decompose at higher temperatures and corrode the surface of the screw and barrel. For working of aggressive materials, the screw and barrel must be coated with protective layers of chromium, nickel, etc.

In conclusion, it may be mentioned that special screws have been developed for specific tasks. Screws for more intensive mixing have another zone added onto the metering zone which may have mixing devices like staggered pins to part the melt in a number of streams which join again. The effect is similar to that of the mixers (see Figs 2.11 and 2.12). A degassing screw, especially meant for hygroscopic thermoplastics, compresses the melt like a conventional screw and then releases the pressure in a subsequent decompression zone with greater depth so that the steam and gases thus released can escape through tiny holes in the barrel. It continues with a second compression zone and a metering zone. It is obvious that the degassing plasticising unit will be almost double in length than a normal one.

The Screw Tip

During injection, when the screw acts as a plunger and pushes the melt through the small orifice of the nozzle, very high forces act on its front part which tend to push the melt backwards. A device called non-return valve or check valve, fixed on the tip of the screw, has the function of allowing the melt to flow in the forward direction during plasticising but prevent its backward flow during injection and the hold-on phase. The most common type of the non-return valve (Fig. 2.14) consists of a sleeve whose external diameter has a sliding fit in the bore of the barrel head. The other part of the valve is an arrow like tip which is fitted onto the front face of the screw through the

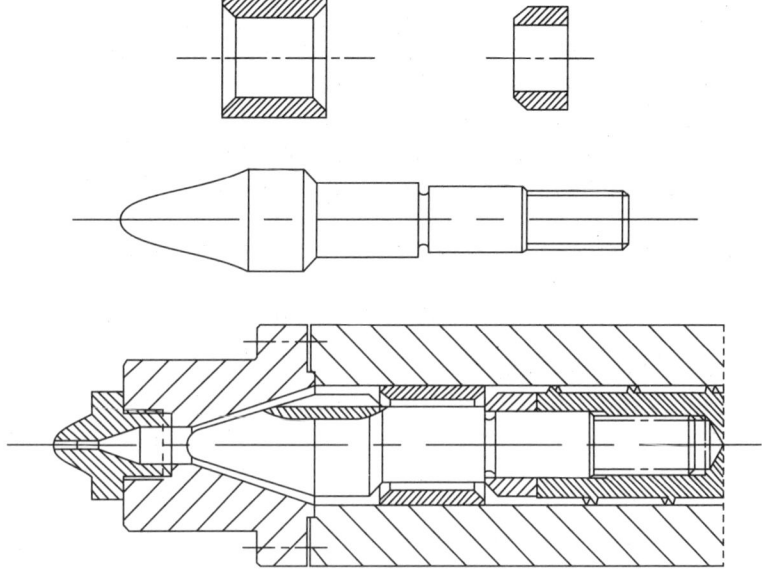

Fig. 2.14

sleeve, leaving a gap for latter's linear movement. The inner bore of the sleeve is larger than the shaft of the arrow head passing through it. The conical arrow head has three or more cut-outs which allow the melt to pass through and collect in the front part of the barrel. The opposite face of the sleeve, when pushed back, sits snugly against the front face of the screw and forms a leak-proof joint.

During plasticising, the melt pushes the sleeve forward and passes through the cut-outs in the tip towards the front part of the barrel, but during injection, the melt in the front pushes back the sleeve against the face of the screw. Now the sleeve forms a piston and prevents passage of the melt backwards.

An improved version of the non-return valve described above integrates a mixing function in the part. The shaft of the tip, which passes through the bore of the sleeve, has shallow spiral grooves cut on it, through which melt can pass forward but partially unmolten granules cannot. Unlike in the first version, the bore of the sleeve here has a sliding fit on the shaft.

Another design of the non-return valve, shown in Fig. 2.15 makes use of a ball in the material channel which allows the melt movement in one direction only.

The functioning of the non-return valves entails passage of the melt through narrow gaps, which may lead to overheating and degeneration of heat sensitive polymers like PVC. For such materials, specially designed screw tips instead of non-return valves are employed (Fig. 2.16). The thread continues, with decreasing depth till the end of the screw tip. Some backwards leakage of the melt is unavoidable but there is no degradation.

Plasticising and Injection

As the screw rotates and transports the material towards the nozzle end plasticising it in the process, it recedes by the same distance to make room for the molten material in the front section of the barrel. However, by throttling the exhaust oil from the injection cylinder at the back of the screw, pressure can be applied and maintained on the melt to increase frictional work and to improve mixing. This is termed as the "back pressure". There is provision for the screw to be pulled back by a short distance at the end of the dosing process called "suck-back" or "decompression". It relieves the pressure on

Fig. 2.15

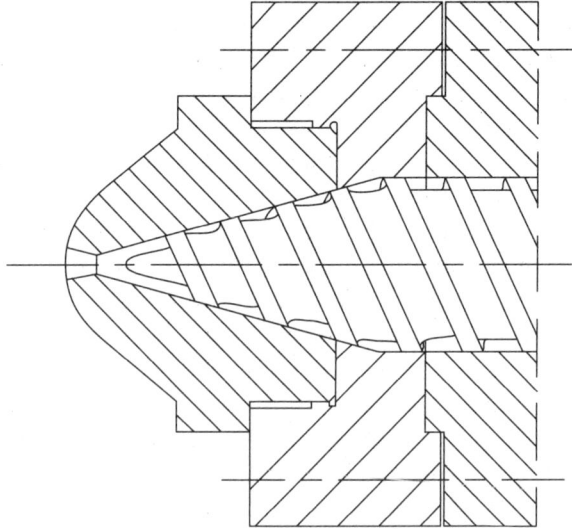

Fig. 2.16

the compressed melt in the front part and thus helps in preventing drooling from the open nozzle. Shut-off nozzles are employed in case of easy flowing polymers, such as nylons.

The heat needed for plasticising the solid material originates from the heaters around the barrel and from the frictional work rendered by the screw. Whereas the heat provided by the electrical heaters is regulated by the thermocouples, that contributed by the screw depends upon its surface speed. Too high a speed may cause overheating leading to degeneration of the polymer.

As described before, the rotating screw makes the granules rub against the barrel wall and form a molten film there. The advancing flank of the thread scrapes the stationary molten material adjacent to the barrel wall and forces it towards the screw root, causing a good mixing and achieving more homogeneity in melt temperature. All material is turned into a viscous fluid before it reaches the end of the compression zone. The back pressure drives any air trapped between the granules back towards the hopper. Finally, the next section, viz. the metering zone, homogenises the melt.

The plasticising unit is capable of linear movement and is held against the mould with sufficient force during injection to prevent leakage. It can be pulled back after dosing to break the contact between the cooled mould and the heated nozzle.

The screw needs rotary movement to act as a transporter and plasticiser. It must also be able to advance axially to inject the melt and to recede during charging to make room for the melt. The rotary motion is imparted by a hydraulic or an electric motor. Both have their advantages and disadvantages. The hydraulic motor has a constant torque but its speed can be varied steplessly. The electric motor is smaller, more efficient and more economical but requires a gear box for speed adjustment which cannot be stepless. In contrast to a hydraulic motor, it has an overload capacity.

A hydraulic cylinder, mounted in line with the screw at the feeding zone end, delivers the force and the linear movement for injection. Another hydraulic cylinder, placed under the unit and coupled with it, moves the plasticising unit linearly towards the mould and away from it. It also delivers the presson force (Fig. 2.17).

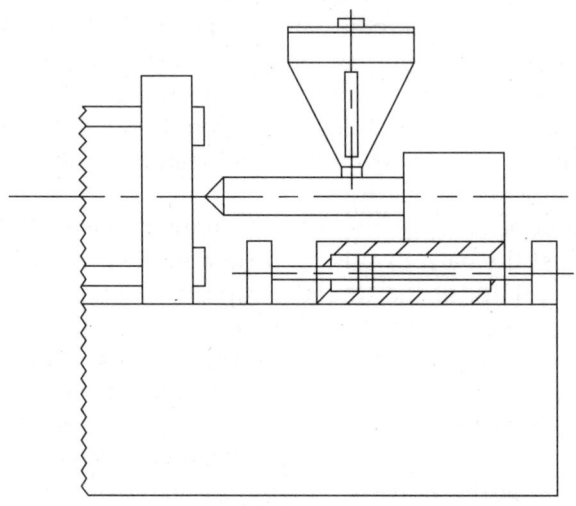

Fig. 2.17

Another movement needed by the plasticising unit is closing and opening of the hydraulic shut-off nozzle. A small hydraulic cylinder performs this operation.

The fully electrical injection moulding machines employ AC-servomotors in conjunction with threaded ball spindles for the rotation and the linear movement of the screw. The force and the movement is transferred with the help of belts made of special elastomers, making the operation smooth and noiseless. The movement of the injection unit too is executed in the similar manner. Figure 2.18 illustrates the arrangement in principle.

The Capacity data of an injection Unit

The parameters which convey the capacity and capability of an injection unit are:

The maximum injection pressure: It depends upon the diameter of the screw. It equals the force from the driving system acting on the screw divided by its cross section.

As a rule, the machine manufacturers offer three different plasticising units for an injection moulding machine. All units have same length but screws of different diameters (and corresponding barrels). Consequently, the specific injection pressures are also different. The biggest screw has the least pressure but more output. It is used for easy flowing polymers. The minimum specific pressure is generally over 1000 bar. The smallest screw offers the maximum injection pressure and is employed for very viscous plastics such as polycarbonate, etc. Its output is the least among the three sets.

The L:D ratio is obviously different in each case.

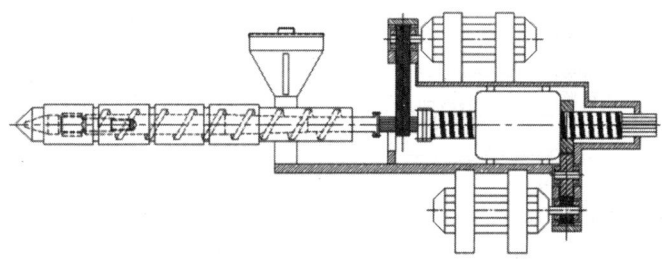

Fig. 2.18

The maximum shot volume: The maximum distance that a screw is designed to recede while dosing is 4–5 times its diameter. In other words, the maximum volume of melt plasticised and transported to the front of the barrel for injection is limited to the cross-sectional area of the bore multiplied by the back stroke. It is, however, advisable to count with about 80% of the rated shot volume. The shot weight will depend upon the specific gravity of the molten polymer.

It may be recalled that all thermoplastics expand when molten. The increase in volume may be as much as 30% or more for semi-crystalline materials and about half of that for the amorphous ones. Consequently, the melt will have a lower specific gravity than that quoted for the solid thermoplastics. The effective shot weight would depend upon both factors, viz. the expansion as well as the specific gravity. Table 2.1 lists the conversion factor for some molten thermoplastics.

Table 2.1: Conversion factors	
Thermoplastics	*Conversion Factor*
PS	0.91
SB	0.88
SAN	0.88
ABS	0.88
CA	1.02
CAB	0.97
PMMA	0.94
PVC-hard	1.12
PVC-soft	1.02
PC	0.97
PE	0.71
PP	0.73
PP+20% Talcum	0.85
PP+40% Talcum	0.98
PP+20% GF	0.85
PA	0.91
POM	1.15

The plasticising capacity: It is a theoretical parameter which states the amount of plasticised material if the screw plasticised continuously. Nevertheless, it furnishes useful information whether the cooling time would suffice for dosing the next shot.

The Mould Closing Unit

Placed in line with the plasticising unit is a set of two vertical plates, also called mould platen (Fig. 2.19). The one on the right, closer to the plasticising unit, is fixed on the machine bed and has a precise round hole in the middle in line with the horizontal axis of the plasticising unit, which allows access to the nozzle of the later and also serves as a locating device for the injection mould. The platen has a number of threaded holes for clamping of one part of the mould (usually called the injection or the fixed mould half). The fixed platen also supports four tie bars (two in case of very small machines), which serve as guides for the other platen, that moves to and fro and carries the other mould half. The mould is generally loaded from above. The larger machines have the provision of pulling out one tie bar in order to load the mould from the side. Like the fixed platen, the moving platen too is provided with threaded holes for clamping of the other half of the mould (ejection or moving half) and has an accurate circular opening in the centre, which provides access to the ejection mechanism and serves as location for the moving mould half.

Fixing of the mould halves on the platen either directly with bolts or with the help of clamps is the classical method widely in practice. Additional devices like mechanical or hydraulic clamps can also be added onto the platen for fast fixing. Some machines are also equipped with electromagnets in the platen

Fig. 2.19

with mechanical safety arrangement so that the mould remains in place even in case of power failure.

The moving platen is moved, by means of hydraulic, mechanical or electrical systems or by their combination, towards the fixed platen to close the mould (i.e. bring both mould halves together) and away from it to open it. The tie bars guide it accurately. In bigger machines, the moving platen is further supported on the bed to prevent bending of the tie bars. Hydraulic or electrical ejection systems are also mounted on the back of the moving platen.

A third plate, parallel to the other two and fixed on the bed behind the moving platen on the same axis, holds the other ends of the tie bars as well as the motive devices such as toggle lever, hydraulic cylinder, etc. In earlier versions, it also held the ejecting rod (see ejection mechanisms).

The latest generation of the injection moulding machines has no tie bars, thus providing better access to the platen, easier loading of the mould horizontally or vertically and better utilisation of the platen area. Guidance to the moving platen provided in an alternative manner, is described later.

The closing unit has the function of closing and opening the mould and holding it closed with a predetermined force during the injection phase. "The clamping force" counters the force, generated by the injection pressure in the mould, which tends to open it. This force is the product of the projected area of moulding and the effective injection pressure per unit area. The maximum area of the article, which can be moulded on a particular machine, can be derived from the clamping force of the machine. It also serves as measure of the machine size.

Summing up, the essential parts of the mould and closing unit of a moulding machine are:

a. A fixed platen (injection side).
b. A moving platen (ejection side).
c. Guiding elements (tie bars and bushes).
d. The third platen.
e. Closing and locking mechanism.
f. Ejection system.

It performs the following functions:
1. Carry the two mould halves.
2. Bring the two mould halves together—precise, parallel and coaxial.
3. Move ejection mould half fast close to the fixed half.
4. Close the mould gently.
5. Build up the locking force.
6. Keep the mould locked during injection and hold-on phase.
7. Release the force after the cooling phase.
8. Open the mould gently till the parting of the two mould halves.
9. Move back fast.
10. Slow down before stopping at the pre-set position.
11. Perform additional functions like ejection, core pulling, etc.

A sliding safety cover with provision to stop any movement of the platen, if it is opened inadvertently, shields the clamping unit.

The Mould Closing Systems

There are three distinct systems of closing (moving the platen) and locking in use. These are:
 I. Mechanical (toggle lever)
 II. Hydraulic (hydraulic cylinders)
 III. Electrical (electrical motors and threaded spindles)

These systems comprise three platen, viz. the fixed mould platen on the injection side, the moving mould platen and a third fixed platen behind the moving one, on which the driving components such as the toggle levers, the hydraulic cylinders or the electrical parts are anchored.

In addition, there are hybrid closing systems in use which combine the above basic types.

Two distinctly different mould closing systems are:
 IV. Two plate locking system (hydraulic)
 V. Tiebarless locking system (hydraulic)

The Mechanical Closing System

The system makes use of toggle lever mechanism to move the middle platen to and fro and to lock it. As shown in Fig. 2.20, it is hinged on one side on the third plate and on the moving platen on the other end. A small, double acting hydraulic cylinder provides the force and the movement to the central hinge of the toggle lever for opening and closing.

The toggle lever closure has proved suitable for injection moulding machines because of its natural attributes, viz.:

- Gentle start, fast movement and then slow movement at the end.
- High closing force at the end of the closing stroke at the cost of very little work.
- Possibility of positive locking by overstepping the dead point.

The toggle lever mechanism may be of the single type as shown above. It is connected to the moving platen in the middle. It is simple in construction but has certain serious drawbacks.

- It does not guarantee absolutely parallel movement.
- It leaves no free space in the centre of the moving platen for placement of the ejection mechanism.
- The position of the hydraulic cylinder is inconvenient. If placed below, it clashes with the machine bed. Fixed upwards, the covers have to be extended.

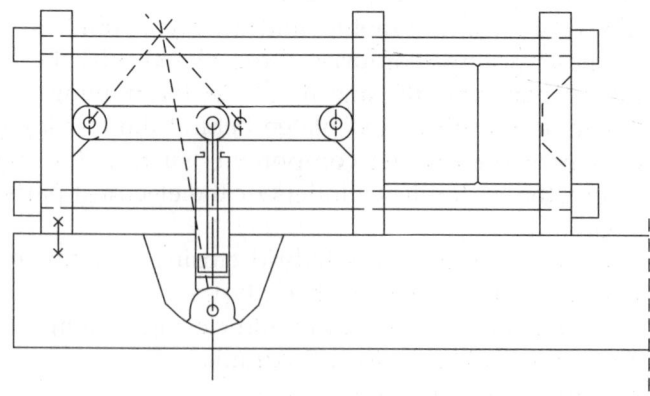

Fig. 2.20

The single toggle lever system, as shown here, is used only with small machines. The system has been refined to a double toggle lever mechanism to eliminate the shortcomings.

The Double Toggle Lever Mechanism (Fig. 2.21)

This is an improvement on the single toggle lever system. It consists of a pair of toggle levers, fixed on one end to the third (stationary) platen and on the other end to the moving platen, on four corners. As shown in the diagram, a hydraulic cylinder placed on the central horizontal axis moves a yoke plate, which in turn transfers the movement to the platen through the toggle levers.

The movement achieved is parallel and uniform. The platen is loaded on four corners. The space in the middle is free for accommodating the ejection mechanism.

Now-a-days, the double toggle lever is used in most of the machines having a mechanical closing system.

The mechanical closing systems generate closing and locking forces by straining two units, viz. the tie bars and the machine platen. When toggle levers make the closing movement and bring the two mould halves in contact, they are not yet fully stretched. The toggle levers can create the locking force only in the fully stretched position. It is achieved by straining all

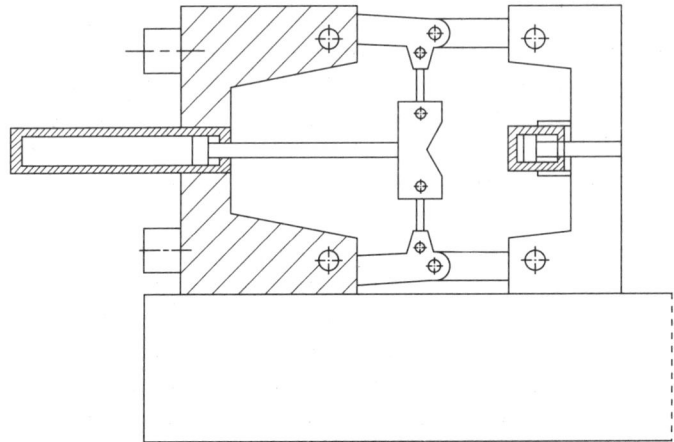

Fig. 2.21

concerned parts, particularly the tie bars. The magnitude of the locking force depends upon the amount of stretching of the tie bars. Should there be any changes in the mould height due to fluctuation of temperatures, the locking force would also not remain constant as the stretching of the tie bars is also effected.

As the stretched length of the toggle levers is unchangeable, adjustments in mould height can be effected only by changing the position of the third plate. Larger machines have the provision to move the plate automatically by means of a motor which operates the four locking nuts holding the plate on tie bars, centrally. The height adjustment in smaller units is carried out manually.

In the case of machines with mechanical closing, the minimum and maximum mould height is specified along with the maximum stroke.

Hydraulic Closing Systems

In fully hydraulic mould closing systems, the force for the movement of platen as well as for locking is brought forth by directly acting hydraulic cylinders. Obviously, cylinders of considerable diameters are needed to generate the required force with the given hydraulic pump pressure. Such big cylinders need enormous quantity of oil for their operation. It will be a waste of energy to execute the long closing and opening movements of the platen with the main cylinder. In any case, the movements would be very slow. Because of these reasons, most hydraulic moulding machines employ a two step closing system:
- Fast cylinder/s with smaller diameter for the movement of platen
- Main cylinder of larger diameter for locking.

A significant difference between the toggle lever and fully hydraulic closing systems is that the hydraulic cylinders can deliver the full locking force at any point of travel whereas the toggles produce it when fully stretched. The force generated by the hydraulic cylinder is directly proportional to the hydraulic pressure from the source. Consequently, it is very simple to adjust its magnitude by adjusting the pressure of the incoming oil. This force remains constant even if the mould height changes due to temperature and also during the injection phase when the pressure of the melt in the mould tends to open the mould. The closing and the locking force are identical in fully hydraulic systems. Non-return valves help to save energy after the required level of locking force has been arrived at.

Unlike the mechanical closing system, the hydraulic closing units have no limitation on the maximum mould height. Their specifications refer to the minimum mould height and the maximum daylight.

Figure 2.22 illustrates the working principle of a fully hydraulic closing system.

Oil is compressible. This is why, hydraulic systems generally employ hydraulic cylinders for the movement and holding of the platen, coupled with mechanical devices for positive locking.

Electrical Closing System

The closing unit of an all-electric moulding machine operates the toggle levers with the help of a threaded ball spindle, which is rotated by a synchronous motor (Fig. 2.23). Though the working is same as with the combination of a hydraulic cylinder and the toggle levers, there are some positive side effects. Oil

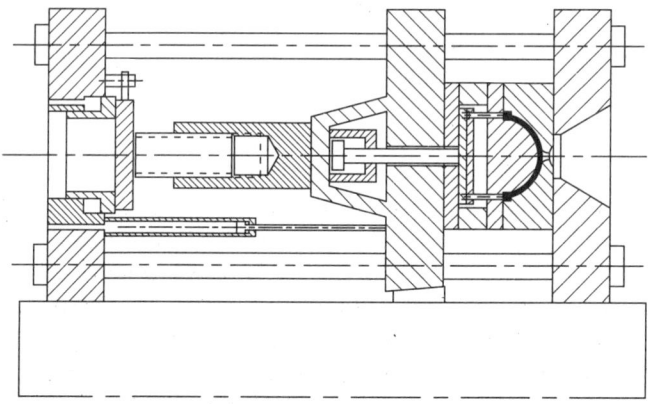

Fig. 2.22

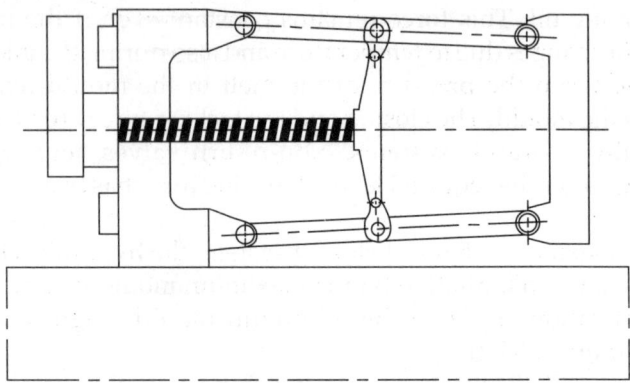

Fig. 2.23

is compressible and the accuracy and consistency of its power and movement transmission is dependent upon its composition, purity and temperature. Leakage also reduces its efficiency. The sizeable volume of oil in large hydraulic machines requires some time to achieve the working temperature. An electrical motive device is free of these shortcomings. The mechanism is clean and drip free and therefore particularly suitable for clean room applications. The machine can be set in operation instantly.

Two Platen Closing System

All foregoing systems employ a third plate in addition to the two mould carrying platen. The third plate is employed for anchoring of the toggle lever mechanism, the hydraulic cylinders or the electrical motor, etc. as otherwise it is not possible to move and lock the mould platen. The same goal can be achieved without the third plate by employing tie bars as piston rods of hydraulic cylinders as shown in Fig. 2.24.

The closing movement is imparted by separate cylinders. The tie bar cylinders take over the locking function. They can be mounted on the back of either the fixed or the moving platen.

The two plate closing system is being increasingly used in small and medium sized injection moulding machines. It is, however, practicable only with the fully hydraulic machines.

The two platen injection moulding machines offer a number of advantages such as:

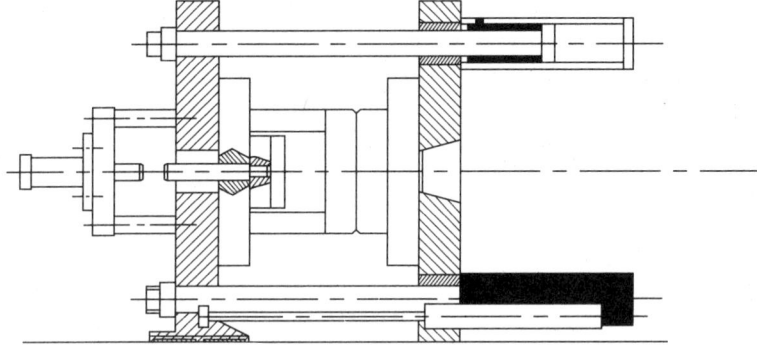

Fig. 2.24

- Less floor area.
- Less weight
- Less oil, saving in energy
- Fast pressure build-up
- Better access to the ejection mechanism
- Fast mould set up
  Here too, a number of variations have been developed.

Tiebarless Closing Systems (Fig. 2.25)
Tie bars have been regarded unavoidable for the movement
and guidance of the moving platen parallel and co-axial to the
fixed platen. They perform this function at the cost of
accessibility and the useful platen area. Loading and unloading
of moulds, which has to be carried out from above, is a time
consuming process. The latest development in closing units of
the injection moulding machines, viz. the tiebarless version,
offers a solution to these very drawbacks.

Engel of Austria has been the pioneer in this field. The idea
has been adapted from the C-presses used for sheet metal
working. The main problem in deleting the tie bars has been to
find an alternative, which would ensure parallelism of the
mould platen, previously taken care of by the tie bars.

A C-frame would obviously open up on the mouth of C when
strained. Incorporation of compensating elastic elements
counters the deformation and keeps the closed mould halves
parallel to each other. It is a complex mechanism.

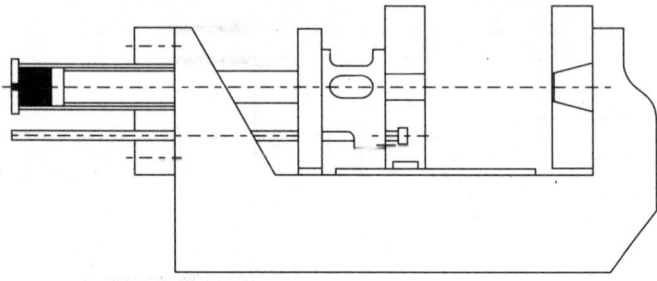

Fig. 2.25

The tiebarless system offers the following advantages over other systems:
- Less cost
- Excellent accessibility from back, front and top for mounting large moulds and for fixtures and handling devices for insert inlaying or part removal.
- Possibility of automatic mould loading and unloading
- Larger useable platen area
- Stiffness of platen without deformation in the middle
- No maintenance of tie bars

Another variation of tiebarless closing systems is an H-frame design.

Comparison of the Toggle Lever and Fully Hydraulic Closing Systems

Toggle levers act on four corners of the moving platen. The injection forces acting on the mould platen through the mould tend to bend them in a concave form. Excessive deformation can lead to flashing. The platen are, therefore, designed to counteract the deforming forces. The permissible deformation ranges from 0.05 mm to 0.3 mm, depending upon the size of the machine.

The hydraulic closing systems apply the locking force in the central region of the moving platen. Consequently, there is no deformation on this side. Excessive injection forces try to bend the fixed platen. The total deformation is, therefore, one sided in case of hydraulic systems. Figure 2.26 illustrates the phenomenon of deformation for both systems.

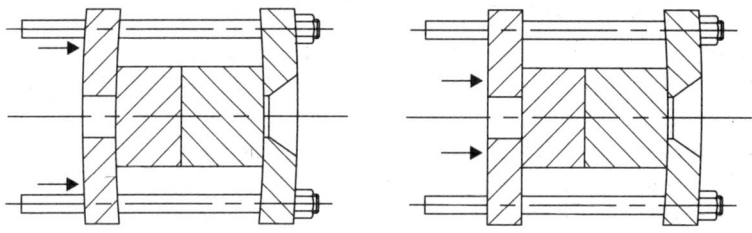

Fig. 2.26

Left. Double Toggle Lever Locking
Right. Hydraulic Locking system

Both systems have advantages and disadvantages:
- The toggle lever system moves the platen ideally without any additional devices. The natural movement of a toggle lever, viz. slow-fast-slow, is what is needed for closing and opening the mould. The hydraulic systems need additional arrangements to achieve the same. Especially in case of big hydraulic machines, with large cylinders and heavy platen, it is a difficult job to accelerate or retard movements accurately.
- Toggle lever systems apply the force away from the centre of the platen where the opposing forces act. The deformation will be more than with hydraulic system acting on the central axis, right in the line of the opposing forces.
- The locking force in case of toggle lever system is dependent upon the extent of stretching of the tie bars. The force changes if the mould temperature and consequently its height changes. The locking force in hydraulic machines is independent of the mould height.
- Toggle lever systems need more space for bigger movements. The machine length increases.
- Toggle lever systems have more moving parts and hence are more expensive to maintain. With hydraulic systems, it is mostly seals and gaskets which need replacement.
- The ejection mechanism can be located conveniently on the central axis with toggle lever operated machines. In case of hydraulic closing systems, special configurations are needed

to create space. More often, alternative systems or locations have to be chosen for the ejection mechanism.

- Hydraulic systems are less noisy than the toggle lever systems.
- The toggle levers develop their locking force in fully stretched state. The minimum and maximum mould heights have to be strictly observed. Hydraulic systems bring forth their locking force at any point of their travel. Only the minimum mould height has to be maintained. There is no bar on the maximum mould height.

The Ejection Mechanism

As a rule, the injection moulds are designed to retain the mouldings on the moving side upon opening. Consequently, the ejection mechanism, which forms a part of the machine, is also housed on the moving side. Its function is to to provide the movement and the force to release the mouldings from the mould parts on which they are holding and to push them out of the mould.

The forces holding the mouldings in the mould depend upon:
- The shrinkage of plastics
- The surface area and thickness of walls in the direction of ejection
- The modulus of elasticity of plastics
- The coefficient of friction between the mould material and the plastics
- Intended or unintended undercuts (caused by roughness of surface and inaccuracies in dimensions)

The ejection mechanism is required to overcome the holding force.

There are three variations of the ejection system.

Mechanical Ejector

The mechanical system of ejection consists of a stationary threaded spindle, housed in the third plate of the machine, in line with the central axis (Fig. 2.27). It forms an obstruction to the ejector bolt of the mould when the moving platen nears the end of the opening stroke. Its protrusion can be adjusted so that the ejector plates of the mould are pushed forward by a

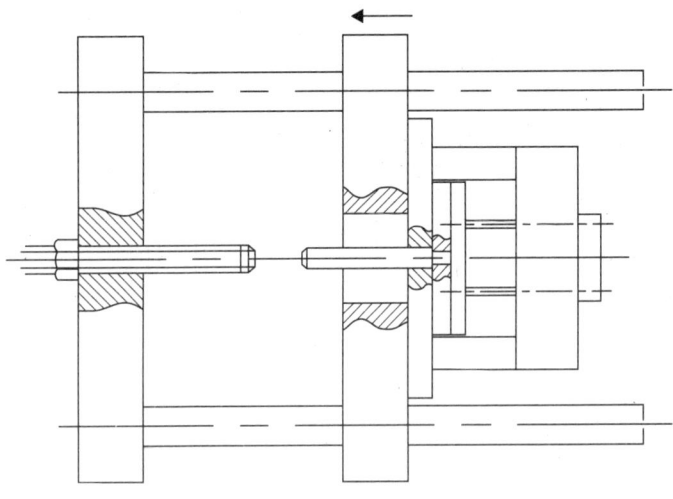

Fig. 2.27

pre-set amount. (Actually, the ejector plates are stopped by the spindle while the mould half moves on backward, thus creating a relative movement of the ejectors.)

The system is very simple but has many drawbacks. It cannot be coupled with the mould; hence it cannot pull back the mould ejector plates. (This may be necessary to place inserts in the mould, to avoid collision of protruding ejector pins with sliders while closing the mould or to enable free fall of the moulding.) The ejector plates of the mould can be actuated only once, which is a distinct disadvantage if the mouldings do not fall down with one push. Also the speed of ejection cannot be adjusted independently.

The ejection force equals the opening force of the machine. The system is almost obsolete.

Hydraulic Ejector

It consists of a hydraulic cylinder, fixed centrally behind the moving platen on the common horizontal axis of the mould closing unit (Fig. 2.28). The cylinder can be coupled with the ejector bolt of the mould. The said cylinder has adjustable stroke, speed and force. It can also be set to pulsate, that is, to

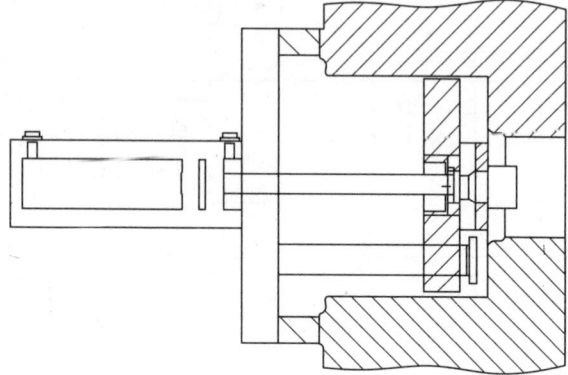

Fig. 2.28

operate a number of times and made to stop in a particular position to enable robots to take over the mouldings. Its start of operation can be initiated through a limit switch or a sensor. In some cases, mostly with large machines, the system comprises an ejector plate, placed directly behind the moving platen and guided by means of guide pillars. It is made to move hydraulically to actuate the ejection system of the mould.

The hydraulic connections of the ejection system can also be utilised to actuate hydraulic motors built in moulds for unscrewing or for actuating slides in moulds through hydraulic cylinders where the core pulling device is not available.

Electro-mechanical Ejector

The fully electrical machines employ a threaded spindle powered by a small electrical servo motor to generate the movement needed to push the ejector plates of the mould (Fig. 2.29). Here too, the stroke, its speed, frequency and position of stopping can be regulated as in the case of a hydraulic ejector.

Pneumatic Ejector

Compressed air is often employed to support other modes of ejection. The moulding machine is equipped with valves to regulate air jets, the start and duration of their operation.

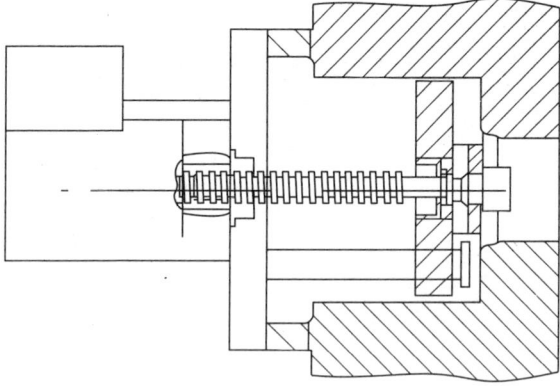

Fig. 2.29

Compressed air at 6–8 bar, generated by a separate compressor, is connected to the machine. The air jets may be incorporated in the mould or placed outside.

The Cooling Battery (Fig. 2.30)

A distribution system for the mould cooling water, having a number of inlets and outlets with valves to regulate the flow, is generally situated at the back of the machine. The common device consists of a main inlet which divides the incoming water¯ into a number of parallel channels. Individual cooling circuits in the mould can be connected to these channels which have transparent sight glasses, floats to indicate the water level and throttles for regulation of the flow quantity. The exit channel is common for all streams. Fresh water may come to the cooling system from a cooling tower or from a refrigeration system. It is used to cool the mould as well as the hydraulic oil tank and the feeding zone of the barrel. One circuit from the cooling arrangement is permanently connected to the feeding section of the barrel of the plasticising unit to prevent melting of raw material close to the entrance, as described earlier.

The used water is generally returned to the initial source through the common exit channel and is cooled again in a closed loop system.

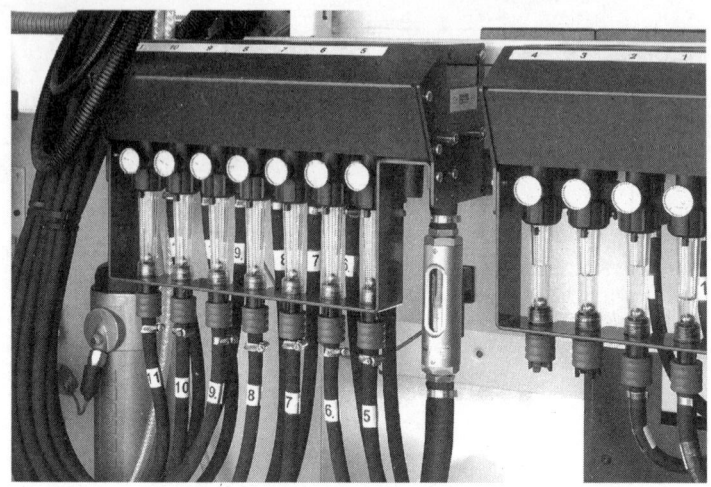

Fig. 2.30

The system, though simple in operation and maintenance, suffers from some severe shortcomings which may effect the quality and productivity adversely. The temperature and the heat conducting capacity of the in-coming coolant depends upon the ambivalent conditions, such as seasonal fluctuations in temperature and humidity as well as from chemical and physical contents of the source water. Depending upon the type of cooling towers, there is some loss of the fluid as well as pollution from surroundings. The simple throttle valves regulate the flow in an unrepeatable way and may often be adjusted without a precise record.

A more reliable system ensuring independence from ambient conditions is the controlled temperature mould cooling gadget, which supplies the coolant at a constant temperature to the mould. It is portable and has a pump, heaters and thermostats and also controlled cooling/heating arrangement for the coolant coming back from the mould. The fluid used is water for lower and oil for higher temperatures. The sophisticated gadgets can supply fluids at temperatures as high as 250°C for processing of engineering polymers.

The Machine Body

The three essential units of the injection moulding machine responsible for processing of the melt described above, are mounted atop a frame which houses the hydraulic system of the machine. It consists of an oil tank, a motor, a pump or pumps, a pressure accumulator, hydraulic valves, oil distribution lines, etc. The pump, driven by the electrical motor, generates pressurised hydraulic oil, which serves as the conveyor of power to actuate various processes such as screw rotation, injection, closing and opening of platen, ejection, etc. by means of suitable devices like hydraulic motors and cylinders. Directional valves direct pressurised oil to various units as commanded by timers.

Hydraulic systems have proved to be most efficient for the purpose of distributing power. Oil is fairly incompressible and obviates the use of rods, levers, racks, cables and gears for transmitting motion. It also permits high switch-over speeds.

The major disadvantage of oil is the change in its viscosity with temperature. Efficient cooling systems and temperature controls can keep this drawback within narrow limits. Leakage is another problem which has been overcome by gaskets and seals made of special synthetic materials, developed specifically for the purpose.

The Control Panel

The moulding operation involves many variables as described before, which must be individually set and controlled for each mould. These are temperatures of various zones of the plasticising unit and sometimes, also of the mould. There are pressures such as injection, holding and back pressure and a number of movements and speeds. The sequence and durations of various steps in the moulding process must be set and maintained precisely. A control panel houses the devices to set and control different parameters like temperatures of different sections of the plasticising unit, speeds of injection, opening and closing of the mould locking unit, injection, holding and back pressures, start and duration of various operations such as injection, holding, cooling, ejection, core pulling devices, etc. Some machines have additional temperature controls for the mould and the hot runners too.

Whereas the older generation of machines operated with limit switches for initiating various operations and ending them, the modern generation uses sensors, which are faster and more accurate. Operations like injection, holding and dosing, screw speed, etc. can be set in a number of steps with changing values to suit the changing conditions during filling, compacting and cooling in the mould.

Most machines are of the controlling, or the open loop type; they bring up the set values of the parameters as accurately as the built-in devices are capable of. The closed loop machines are equipped to control certain parameters and correct them if necessary. For example, if the set speed of injection is not being maintained due to some change, say due to variation in the viscosity of the material caused by a new charge, the oil input to the injection cylinder is automatically regulated in order to correct the speed. The parameters controlled and regulated may be the injection speed, the pressure, the thickness of the melt cushion in front of the screw tip, etc. depending upon the design of the machine. The closed loop machines contain one or more microprocessor-based computers integrated in the system which receive signals of various parameters such as screw position, screw speed, holding pressure and other chosen parameters, process the signals and generate specific output for altering some functions such as opening or closing valves, changing pressure, increasing or decreasing temperature. The complete data can be stored and reactivated as soon as the same mould is loaded again.

The closed loop system contributes to the uniformity of production, narrow dimensional variation and low rejection by keeping the moulding conditions within given limits.

SPECIAL MACHINE CONFIGURATIONS

Injection moulding machine with a horizontal injection unit and a horizontal mould closing system may be considered as a norm. However, these two main units may also be placed differently for special applications.

Vertical Injection; Horizontal Locking

The combination of horizontal locking with vertical injection unit is employed when the injection should take place on the parting line of the mould. It is often the case with 2-component moulding.

Vertical Closing; Horizontal Injection

This configuration is useful for insert moulding. Insert to be moulded-in can be easily placed in the lower half of the mould fixed on the lower, immovable machine platen without additional devices to hold the inserts in place. The upper platen moves up and down to open and close the mould. The injection takes place on the parting line.

Alternatively, both units may be situated in the vertical position, if injection in the centre of the mould is more advantageous.

Vertical Closing with Sliding Lower Platen; Vertical Injection

The combination is especially employed for insert moulding. The mould has one upper half but two identical lower halves, fixed on the sliding table. When injection is taking place with one of the lower halves in engagement, inserts can be loaded in the other half simultaneously. The lower platen slides toward right or left alternatively after the mould opens. The loaded part of the sliding platen comes to the injection position and the moulded parts can be ejected from the other station. It can be reloaded with inserts for the next cycle.

The injection unit can also be mounted in horizontal position for injection on the parting line.

Vertical Closing with Revolving Lower Platen and Vertical Injection

The combination is similar to the foregoing one. The lower platen consists of a revolving table with 2, 3 or 4 stations. Accordingly, the number of identical lower mould halves corresponds to the number of stations. The single upper mould half is common to all. The function is similar to the configuration with sliding lower platen. However, ejection and reloading can be carried out separately with three stations. An additional station offers the possibility of carrying out some after-operation like testing.

Here too, the injection unit may be placed horizontally. Figure 2.31 shows a 2-station revolving table moulding machine with vertical injection.

SPECIFICATIONS OF INJECTION MOULDING MACHINES.

Machine specifications are technical data which give information about the type, capacity, input and output, effeciency, accuracy, space requirement, maintenance, etc. The information forms the basis for selecting a suitable machine for a present job or future project.

The most vital information is conveyed by the machine designation as formulated by Euromap, the European committee of manufacturers of plastics machinery. For example, an injection moulding machine with a locking force of 1500 kN, 500 CC swept volume and a horizontal locking unit would be designated, according to Euromap guidelines, as 1500 H-500.

Fig. 2.31

The other machine specifications needed by mould designers and moulders are:

- The injection rate
- Maximum injection pressure
- Platten size, horizontal and verticle distance between tie bars
- Maximum and minimum mould height and opening stroke or
- Maximum daylight and minimum mould height
- Ejection stroke and ejection force
- Dry cycle time
- Energy requirement-total connected load
- Net weight and floor space requirement.
- Optional equipment such as plasticising unit for PVC, degasing injection unit, shut off nozzle, core pulling device, handling device, etc.

The detailed specifications include more information about the plasticising unit, the screw, the mould closing and locking system, the ejection system, the range of optional equipment, etc. which is needed when several machines have to be compared before making the final selection.

3

Injection Moulding of Thermoplastics, the Process

- *The process*
- *The screw pre-plasticisation*
- *The mould filling process*
- *Post operations*
- *Trouble shooting*

THE PROCESS

The injection moulding process comprises three distinct steps, viz. plasticising the solid plastics raw material, injecting it into a closed mould having the cavity/cavities corresponding to the product to be moulded, cooling and solidifying it there before extracting the mouldings after opening the mould. The set of these sequences is termed as the moulding cycle.

The Moulding Cycle

The moulding cycle begins when the moving platen gets the command to close the mould. The closing movement is executed in three steps, viz. slow, fast and again slow at the end. The locking force is built up.

The second step in the cycle is bringing the injection unit forward till the machine nozzle comes in contact with the sprue bush of the mould and presses on it.

The next operation is the injection of the melt. It takes place in three stages. The first stage is the complete filling of the mould through the linear stroke of the screw acting as a piston. In order to force the melt in the cavity to fill it, pressure is needed to overcome the resistance to flow in the nozzle and the mould. The second stage comprising compression of injected melt follows immediately thereafter. Plastics melts are compressible and the screw pushes in more material but with higher pressure for compaction. A melt cushion is left in front of the screw tip.

The compression phase is followed by the holding phase, which forms the third stage. The holding pressure, usually lower than the last one, pushes in melt from the cushion to make up for the shrinkage taking place during solidification of the melt till the gates are frozen.

At the termination of the holding period, the cooling period starts. It constitutes the longest part of a moulding cycle. Its duration is chosen in relation with the thickness of the moulding. The dosing for the next cycle, decompression and pulling back of the injection unit is also carried out during this phase (the injection unit is pulled back so that the heated machine nozzle does not lose heat to the relatively cold mould sprue bush).

The mould is opened as soon as the preset cooling time is over and ejection takes place just before the moving mould half reaches its end position.

Summing up, the complete moulding cycle, yielding one shot, consists of following segments in the given sequence:

1. Closing time
2. Nozzle forward time
3. Injection time
4. Holding time
5. Cooling time
6. Opening and ejection time

Closing Time

It is the time needed by the moving platen to move forward and close the mould. The major part of the stroke is covered very fast but the platen slows down before touching of the two mould halves so that the final closing and locking takes place gently.

Nozzle Forward Time

The retracted plasticising unit moves forward till it touches the closed mould. The machine nozzle presses against the sprue bush of the mould with certain force to prevent leakage during injection. The moulding operation can begin.

Injection Time

The time needed by the plunger to force the melt in the cavity/cavities of the closed mould is termed as the injection time. It also includes the time for compression of the melt after filling. Usually, the injection time forms a small fraction of the total cycle as it is desirable to fill the mould as fast as possible to have the plastics temperature nearly same in all parts of the cavity and also to prevent premature freezing of the melt.

Holding Time

After injection, the melt starts cooling and contracting. In order to compensate for the space created by contraction, the pressure is kept on by the plunger on the remaining melt cushion in the front part of the barrel till all melt in the cavity is frozen (the gate should also freeze at this juncture). This step counters the formation of sink marks and voids. The holding pressure is usually lower than the injection pressure but its duration may be much longer. The switch-over from injection to the holding phase is triggered by a time relay or by a distance switch or by a hydraulic pressure sensor. It can also be initiated by a pressure sensor built in the injection mould which effects the switch-over when a preset pressure has been arrived at in the mould.

Cooling Time

The so-called cooling time starts after the holding time is over. It constitutes the longest part of the total moulding cycle. The moulding is allowed to dissipate its heat through cooled mould walls till it cools down to a temperature where it can be ejected without distortion. Main part of the shrinkage takes place during this phase.

Opening and Ejection Time

At the end of the cooling time, the moving platen is pulled back in the same manner as during closing, viz. slow, fast, slow and the mould is opened. Towards the end of the opening stroke, the ejector is activated and the cooled mouldings are pushed out of the moving half of the mould.

The recharging or dosing of the cylinder for the next shot takes place during the cooling time. Usually, the plasticising unit is retracted after recharging so that the nozzle of the barrel does not lose heat to the cold mould.

The recharging time, depending upon the shot weight, can be influenced by the screw speed.

Out of the six time segments forming the cycle time, three segments, viz. mould closing, injection unit forward and opening cum ejection are dependent on the machine and can be varied only marginally. Injection as well as holding time are directly linked with the polymer, the size of the moulding and its wall thickness. These time segments, too, are quite short as it is aimed to fill the cavity as fast as possible. It is the cooling time which forms the major segment of the cycle, coming up to 80% of the total time for one shot. It can be influenced favourably through a good mould design. The cycle is generally portrayed as a full circle. Figure 3.1 portrays relationship of the various segments of a cycle in visual form.

The exact duration of a moulding cycle can be determined only after trials and optimisation of the process. However, for rough estimation of the cycle time for a planned product with normal wall thickness, the following thumb rule may be found helpful.

Cycle time = 2.5x (max. wall thickness)^2 seconds

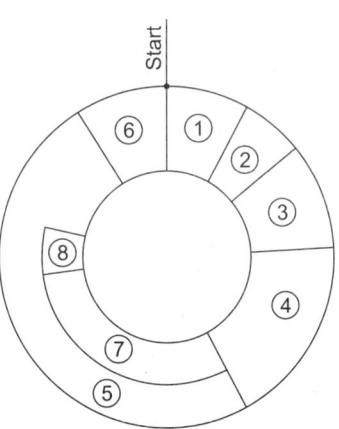

① Mould closing
② Injection unit forward
③ Injection
④ Holding
⑤ Cooling
⑥ Opening + Ejection
⑦ Dosing
⑧ Injection unit retraction

Fig. 3.1

THE SCREW PRE-PLASTICATION

The rotating screw transports the material from the feed zone towards the delivery end over the compression and the metering zones. In the process, the material is compacted and made to rub against the inner wall of the barrel which brings it to melting and sticking to the inner wall of the cylinder. The advancing flank of the rotating screw scrapes it free and sends it inwards in a curling movement. Its place is taken by unmolten material. The continuous process of melting, scraping and replacement results in mixing and homogenising. The metering zone with its constant depth stabilises the melt before it goes through the non-return valve and collects in the front part of the barrel vacated by the gradually receding screw. The amount of the material plasticised and forwarded is directly proportional to the speed of rotation.

The screw speed influences the time needed for dosing the required amount of melt for the next shot but also the friction generated. It is usually adjusted to accomplish the first task within the set cooling time. Too high a speed can lead to overheating and decomposition of the polymer. As a thumb rule, the surface speed should not exceed 15 meters per minute for the thermally less sensitive commodity plastics. It should be lower for technical polymers. Table 3.1 contains the recommended surface speed for different thermoplastics.

Table 3.1: Surface speed of the screw	
Plastics	*Surface speed meters per min.*
Polyolefins	15
Polystyrene	15
SAN, ABS,	6
ABS-PC-Blend	4
PVC-rigid	2
PVC-plasticised	6
POM	7
PET, PBT	7
PMMA	6
Polycarbonate	6
PPO	6
Polyamides	10

As the screw recedes continuously during plasticising, its effective length over which the material travels forward and is subjected to friction, also decreases. This results in gradual decrease in temperature of the melt. It can be compensated, either by raising the speed of the screw continuously or by increasing the back pressure while keeping the speed constant.

The amount of dosing should consist of the melt needed for filling, compacting and compensating shrinkage as well as an additional amount as cushion to prevent the screw tip hitting the end of barrel. The melt cushion may vary between 5-10 mm. in thickness.

THE MOULD FILLING MECHANISM

When hot melt is injected in the relatively colder mould, the melt layer coming in contact with its wall loses heat and comes to a halt. The layer adjacent to the frozen layer experiences maximum friction and flows with reduced speed. The layer next to it has less friction and higher velocity which in turn reduces the viscosity. The layer in the centre of the stream has maximum speed and also maximum temperature. The flow front advances parabolically (Fig. 3.2). The pattern is called

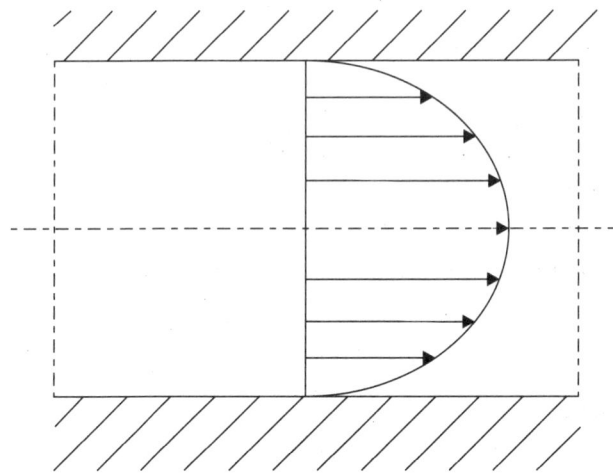

Fig. 3.2

fountain flow with all layers having different temperature, different viscosity and different speed. As the flow proceeds, the melt from inner layers moves forward as well as upwards, touches the mould walls and comes to a halt.

The speed with which the melt proceeds in the mould cavity, depends not only on the pressure exerted on it by the ram but also on the brake applied by the cooling which again depends upon the mould temperature, the heat conductivity of the tool steel, the temperature of the melt and its speed and the corresponding viscosity. The shear caused by the flow generates heat, lowers viscosity and influences the speed of the melt. This is why, a decrease in pressure results in a disproportional decrease in velocity of the melt. The decreased speed accelerates the loss of heat and increases viscosity which means less heat generation through shear and further rise in viscosity and in loss of speed.

The injection part of the cycle also includes compaction at its conclusion. The plunger moves at a very slow speed and compacts the melt. It is followed by the holding phase which takes place at comparatively lower pressure but for a longer period to compensate the contraction caused by shrinkage during cooling. It comes to an end when the gate gets frozen and no more melt can be forced into the cavity irrespective of the pressure. It signals the end of the mould filling phase. Figure 3.3 illustrates this graphically.

The complex relationship between pressure, velocity, temperature, heat conduction, viscosity and heat generation through shear makes it difficult to predict the optimal moulding parameters in advance. The optimum cycle is always determined through trials. Softwares like moldflow and many more take these factors into consideration and calculate the parameters with fair accuracy if the relevant characteristics of the particular material grade are known.

THE FORCES

Various forces involved in the process of injection moulding are:

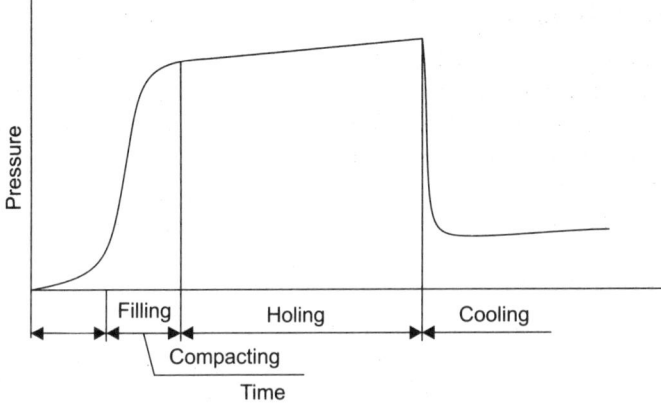

Fig. 3.3

Injection Pressure

It is the force with which the screw, acting as a plunger, is pushed forward to force the melt into the mould. Its magnitude is decided by the extent of resistance to the flow of the melt in the machine nozzle, in the mould feeding system and in the mould cavity, which in turn depends upon the wall thickness of the moulding, the viscosity of the melt, the speed of injection and the temperature of the mould.

In modern machines, injection speed is the parameter set in the place of pressure. The aim is to maintain an even flow of melt in the mould. Although it is desirable to fill the cavity as fast as possible so that the temperature of the melt is almost same in all parts of the cavity, it is recommended to start with a lower speed in the beginning to avoid turbulence and then increase the speed, cutting it down again at the end to prevent flashing.

The pressure set on the machine indicates the upper limit which may be reached for maintaining the speed of injection.

The modern moulding machines have the provision to set the moulding pressure alternatively the injection speed in a number of variable steps. The aim is to maintain a uniform speed of the filling front in the mould.

Holding Pressure

Holding pressure, as described before, has the function to force additional melt into the mould to compensate for the reduction in volume of the melt through shrinkage while solidifying. It remains active till the gate freezes.

The holding pressure can be selected in variable steps compatible with the rate of cooling in modern machines.

The Back Pressure

The back pressure acts against the free backward movement of the screw during recharging and helps to subject the melt being charged to greater friction. It is much lower than the injection or holding pressure.

The closed loop machines employ back pressure to regulate friction which generates heat for melting of the plastics. As the screw recedes during charging, its length over which the subsequent material travels (from hopper till screw tip) is reduced. In other words, the plastics being worked upon at this stage undergoes less amount of friction and has consequently lower heat input. It can be compensated by increasing the back pressure in steps.

The Locking Force

The locking force of the mould closing unit counters the force created by the pressurised melt in the mould, which tends to force it open. It is proportional to the effective injection pressure and the projected area of the moulding. It is a guide to determination of the size of the moulding machine for a particular mould. The locking force should exceed the expected force generated by the injection pressure in the mould by about 20%.

The Nozzle Lay-on Pressure

The nozzle lay-on pressure serves to neutralise the backward pressure exerted by the injected melt upon the plasticising unit through the sprue. It depends upon the size of the sprue and the extent of the injection pressure. Insufficient nozzle lay-on pressure would result in leakage of the melt at the junction of the machine nozzle and the sprue bush of the mould.

The Ejection Force

The ejection force, variable in modern machines, has to overcome the forces with which the mouldings stick on the moving side of the mould. These forces are given rise to by the shrinkage, condition of the mould surface, friction, undercuts, etc.

POST OPERATIONS ON MOULDINGS

It has been observed in the outset that injection moulded articles are generally ready for use right after ejection and cooling to the room temperature. The exception to this rule is formed by parts moulded of hygroscopic thermoplastics. The raw material is dried before moulding and the mouldings are free of moisture. However, they achieve equilibrium, their normal physical properties and final dimensions only after absorbing moisture. In most cases, it takes place automatically if the moulded parts are stored for some time. Polyamides, however, need a very long period and proper ambient conditions to achieve equilibrium. Under normal atmospheric conditions, a PA 6 moulding with 2 mm. wall thickness may take as many as ten weeks to absorb the normal amount of moisture, that is 3% by weight and have its normal dimensions and other properties. The dimensional increase after saturation may be about 0.8%.

Conditioning

Polyamide mouldings are allowed to cool down for a day, preferably in the moulding shop. To expedite the process of moisture absorption, the polyamide mouldings are stored in water for a duration depending upon their wall thickness. The duration can be shortened significantly by increasing and maintaining the temperature of the conditioning water bath around 60°C. A 3 mm. thick PA 6 moulding will require 30 hours in a 40°C water bath but only 2 hours if the water temperature is raised to 80°C. As most mouldings have thick and thin sections, the most suitable duration can be ascertained through trials.

The mouldings should be packed in polyethylene bags and stored for sometime and then post-conditioned at room temperature in humid surroundings. It may take a few days to a few months, depending upon the wall thickness. The mouldings achieve a stable state and maintain their properties.

The raw material manufacturers generally recommend the appropriate temperatures and durations for their products.

TROUBLE SHOOTING

Moulding Faults, their Causes and Remedies

The process of injection moulding involves temperature, pressure, time and speed. Most moulding defects are caused by the wrong use of these factors and can be remedied by the proper application of the same during the moulding process. Some defects, however, are the outcome of unbalanced article and/ or mould design. The corrective measures for these faults may be tedious and expensive.

Short Shots

Short shot or an incompletely filled moulding is the result of molten material solidifying before the cavity is fully filled. The defect appears in areas away from the gate, near thin sections and in deep recesses.

Causes

a. Melt temperature too low
b. Mould temperature too low
c. Injection pressure too low
d. Injection speed too low
e. Insufficient injection time
f. Insufficient feed
g. Air trapped in the cavity
h. Too small a gate
i. Wrongly located gate

j. Cold slug blocking the flow
k. Switch-over from injection to holding pressure too early

Remedies

1. Increase the melt temperature
2. Increase the mould temperature
3. Increase the injection pressure
4. Increase the injection speed
5. Increase the injection time
6. Increase the feed
7. Improve venting
8. Increase the gate/runner size
9. Change the gate location
10. Enlarge the cold slug trap
11. Delay switch-over to holding pressure

Flash

Flash is the undesirable extra material appearing as a thin film at the parting line, in holes and apertures, around inserts and ejectors, which necessitates additional post operations for removal.

Causes

a. Badly matched parting line
b. Too much clearance between apertures and inserts, holes and pins, holes and ejectors
c. Vents too deep
d. Elastic warpage in mould plates
e. Insufficient clamping force
f. Very high injection pressure
g. Melt temperature too high
h. Switch-over to holding pressure too late
i. Unbalanced runner system (multi-point/multi-cavity gating/family mould)

Remedies

1. Improve mould workmanship
2. Strengthen mould bolster

3. Reduce melt temperature
4. Reduce moulding pressure
5. Switch over to holding pressure earlier
6. Balance the runner system
7. Change over to a higher capacity machine
8. Replace worn out ejectors, core pins and loose inserts

Difficult Demoulding

The moulding sticks rigidly to core or cavity and cannot be ejected without distortion or damage. It may also stick on the side without ejection arrangement.

Causes

a. Bad workmanship, rough mould surface
b. Highly polished surface (in case of some plastics like PP, PVC, etc.)
c. Insufficient draft angle
d. Poor venting
e. Overpacking
f. Unbalanced/unsuitable ejection system
g. Moulding too hot

Remedies

1. Improve mould surface, remove undercuts
2. Sand blast the mould surface (PP, etc.)
3. Increase demoulding taper
4. Improve venting
5. Reduce holding pressure/time
6. Redesign ejection system
7. Increase cooling time
8. Use differential cooling
9. Use mould release agents

Dimensional variations

The dimensions of the mouldings are not constant.

Causes

a. Moulding conditions not constant
b. Mould temperature not constant

c. Non-return valve leaking
d. Cooling time insufficient
e. Oil temperature varying

Remedies

1. Check thermostats
2. Cool mould through a controlled water temperature gadget
3. Check the thickness of melt cushion and if varying, change valve
4. Increase cooling time
5. Check and clean oil cooling system.

Sink Marks, Voids (Fig. 3.4)

Sink marks are shallow depressions on the surface of the moulding and voids are hollow areas inside a moulding. They are different forms of the same defect. The contraction in volume during cooling has not been compensated and makes its appearance as a sink mark. If the mould is very cold, the upper layer may freeze soon and become hard, transferring the defect inwards as a void.

Causes

a. Uneven cross section, thick and thin areas, material accumulation

Fig. 3.4

b. Unsuitable gate location
c. Insufficient gate/runner size
d. Mould too cold
e. Insufficient holding pressure/time
f. Insufficient material cushion in the cylinder
g. Too early switch-over to holding pressure

Remedies

1. Redesign the article
2. Increase gate/runner size
3. Place the gate close to the thick section
4. Increase mould temperature
5. Increase holding pressure/time
6. Increase material cushion
7. Delay switchover

Weld Lines

A weld line is the meeting of two melt streams and appears as a thin hairline if the streams rejoin properly. Under unfavourable conditions, it may look like a notch. In most cases, it is only an aesthetic shortcoming. However, in case of filled plastics it does constitute a weak spot, prone to failure due to orientation of the fibres.

Causes

a. Parting of melt due to obstruction in the flow path, such as core pins for holes
b. Multiple gating
c. Uneven wall thickness
d. Unequal flow lengths of material streams
e. Trapped air between the flow fronts

Remedies

1. Change gate location to shift the weld line to a less critical area or closer to the gate.
2. Increase the size of gate and runners
3. Raise the melt temperature to make the weld stronger.
4. Raise the mould temperature
5. Increase the injection speed to prevent cooling of flow fronts

6. Improve venting
7. Increase the holding pressure/time

Record Effect (Fig. 3.5)

Fine, concentric lines on the surface of the moulding are termed as gramophone record effect.

Causes

a. High resistance to melt flow in the cavity (slip stick effect)
b. Mould too cold
c. Material temperature too low
d. Injection speed too low

Remedies

1. Increase mould temperature
2. Increase material temperature
3. Increase injection pressure/speed

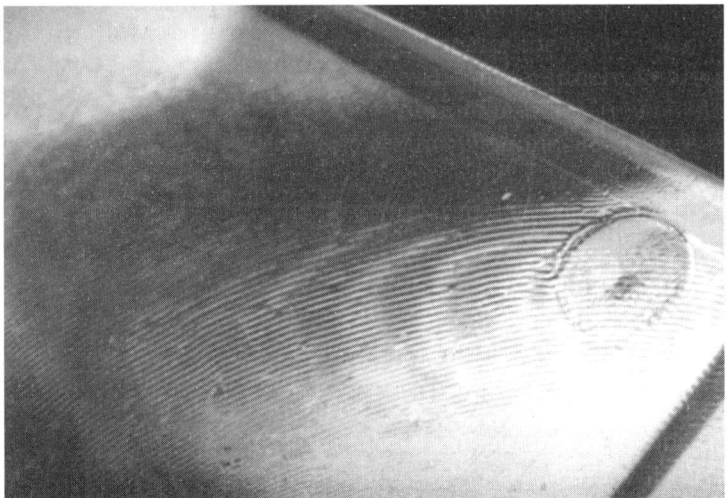

Fig. 3.5

Jetting (Fig. 3.6)

Jetting has the appearance of a worm, usually starting from the gate and extending towards the farthest point of filling.

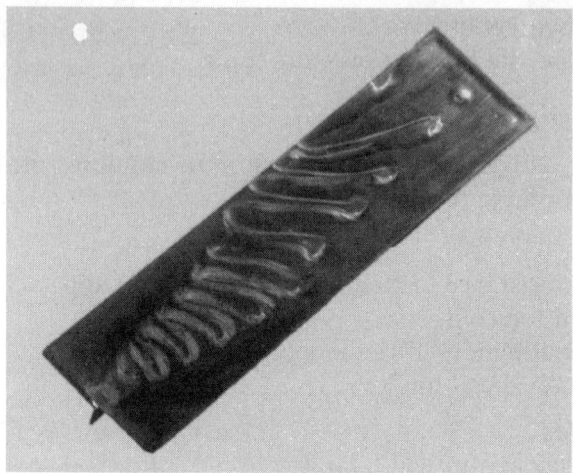

Fig. 3.6

Causes

a. Wrong gate location causing free flow. Melt travels unhindered to the distant wall of the cavity
b. Melt temperature too low
c. Gate size too small

Remedies

1. Change the gate location to make the melt stream meet a wall/hindrance immediately
2. Increase the melt temperature
3. Increase the gate size
4. Increase the mould temperature
5. Reduce injection speed/change injection profile (slow-fast)

Burn Marks

Burn marks occur when the melt gets charred due to abnormally high temperature rise. It happens when the air in the mould gets trapped and is compressed either between two material streams, at the end of the flow path or in blind holes/depressions such as ribs and bosses. It is also called the "Diesel Effect".

Causes

a. Wrong gate location
b. Poor venting
c. Uneven wall thickness
d. Very high filling rate
e. Overheating of the melt

Remedies

1. Relocate gate to eliminate air entrapment
2. Improve venting
3. Change wall thickness
4. Decrease injection speed
5. Decrease material temperature

Burn Streaks at Gate (and/or surface)

These are discoloured rays, emanating from the gate or sprue.

Causes

a. Material temperature too high
b. Dwell time in injection unit too long
c. Gate too small
d. Injection speed too high

Remedies

1. Reduce material temperature
2. Switch over to a smaller machine
3. Increase the gate size
4. Decrease the rate of injection

Silver Streaks

Silver streaks are elongated bubbles pulled in direction of flow and appear like shining rays emanating from the gate.

Causes

a. Moisture in granules
b. Water leakage in mould
c. Melt overheated and decomposing
d. Air not expelled from melt
e. Air being sucked in

Remedies

1. Pre-dry the material
2. Use heaters in the hopper
3. Decrease melt temperature
4. Decrease injection speed
5. Increase back pressure
6. Reduce retraction speed of screw while decompressing

Fish Eyes

It is a defect caused by unmelted or partially melted granules in the melt stream which appear on the surface of the moulding.

Causes

1. Low barrel temperature
2. Low screw rpm
3. Low back pressure
4. Contamination in material, mixing with high temperature plastics
5. High proportion of regrind

Remedies

1. Increase barrel temperature
2. Increase screw rpm
3. Increase back pressure
4. Clean injection unit, charge new material; store materials separately
5. Reduce proportion of reground material

Orange Peel

This defect manifests itself as an uneven, pit-marked and sometimes, as a dull surface.

Causes

a. Material temperature too low
b. Mould temperature too low
c. Injection pressure too low
d. Injection time too low
e. Holding time too low

Remedies

1. Increase the material temperature
2. Increase the mould temperature
3. Increase the injection pressure
4. Increase the injection time
5. Increase the holding time.

Blushing

Blushing is the dull surface around the gate.

Causes

a. Cold material from the nozzle
b. Mould too cold

Remedies

1. Raise the nozzle temperature
2. Prevent drooling, employ suck back
3. Pull back injection unit after dosing
4. Provide/enlarge cold slug well

Delamination (Fig. 3.7)

The moulding is formed of layers, which can be pulled apart.

Fig. 3.7

Causes

a. Mould too cold
b. Material contaminated, mixture with incompatible material
c. Material too cold
d. Moisture in the material
e. Colorant incompatible
f. Injection speed too low

Remedies

1. Raise the mould temperature
2. Purge the barrel
3. Pre-dry the material
4. Change dye/pigment
5. Increase the rate of injection

Warpage

Warpage is distortion of the product surface, caving in of the walls, twisting and waviness.

Causes

a. Wrong product design, large flat surfaces, unequal wall thickness
b. Non-uniform cooling of the mould
c. Too short a cooling time
d. Holding time too short
e. Unbalanced ejection forces
f. Wrong location of the gate

Remedies

1. Improve product design
2. Balance mould cooling
3. Increase cooling time
4. Increase holding time
5. Increase demoulding taper, improve surface finish of the side walls.
6. Add ejectors at crucial points
7. Relocate the gate
8. Change to an amorphous material if possible.

Surface Gloss not Uniform

Shiny product surface with dull patches.

Causes

a. Mould temperature not uniform
b. Mould polish not uniform
c. Materials of mould components have different heat conductivities
d. Wall thickness differences
e. Sink marks
f. Cold material

Remedies

1. Improve mould cooling
2. Repolish mould surface
3. Redesign product
4. Heat up mould before production start
5. Increase holding time/pressure
6. Increase nozzle temperature
7. Pull back injection unit after recharging
8. Enlarge cold slug well

Blistering

Bubbles containing air, visible in transparent parts or thin walled mouldings. They may also appear as projections (different from voids).

Causes

a. Entrapped air
b. Drawn-in air
c. Gases from decomposed melt

Remedies

1. Improve venting; relocate vents
2. Decrease melt temperature
3. Decrease dwell time; decrease cooling period
4. Decrease screw rpm
5. Decrease suck back
6. Enlarge nozzle bore

Design of Injection Moulds for Thermoplastics

- Design of Injection moulds
- Special injection moulds
- Classification of Injection moulds
- Procedure for designing injection moulds
- Materials for injection moulds
- Manufacturing techniques for injection moulds
- Design of injection moulded articles

DESIGN OF INJECTION MOULDS

Injection moulding is one of the major processes for producing articles out of thermoplastics. The injection moulding machine performs the process with the help of an injection mould. It is the injection mould which converts the molten plastics supplied by the machine into the intended article and to be able to do this, it must have a hollow space corresponding to the article. The "hollow space" or the mould cavity is formed by two mating components called "cavity" and the "core". These constitute the basic mould. The cavity is generally situated in the fixed half of the mould and the core in the moving half. The mating surface of the two mould halves is referred to as the "parting line".

The cavity or the outer contour of the product may be machined directly in the solid plate forming the fixed mould half or in a separate block, inserted in that plate. It is then referred to as the "cavity insert". Depending upon the shape of the product, the insert may be round, square or rectangular as shown in the introductory design (Fig. 4.1). Similarly, the core forming the inside of the moulding may be shaped out of the solid plate or inserted into the core plate. In case of multi-cavity moulds, the cores are invariably inserted.

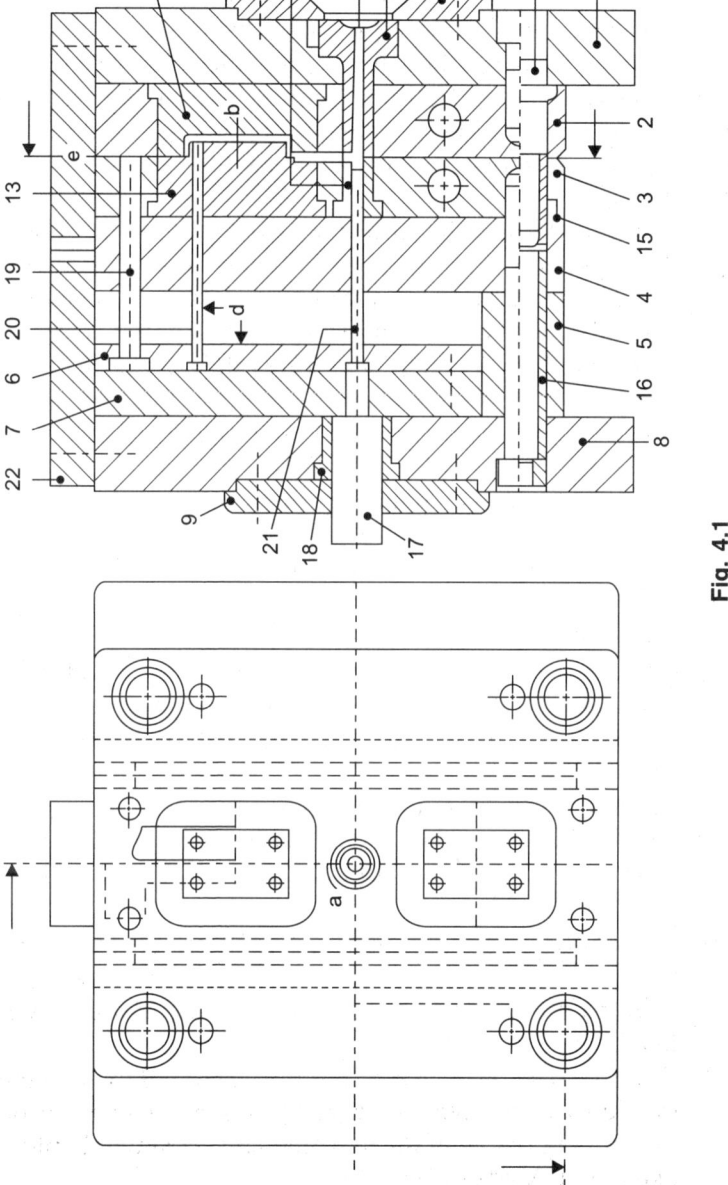

Fig. 4.1

An injection mould has to perform three main functions:
- Feeding the injected melt into the mould cavity.
- Cooling the injected melt to a rigid article.
- Ejecting the rigid moulding without distortion.

In order to discharge these functions, a mould must contain, besides the core and cavity, three facilities, viz. feeding, cooling and ejection systems. These systems are put into practice with the help of a number of components which constitute the mould.

The mould design shown in Fig. 4.1 depicts the essential mould components, their function and their inter-relationship. It is intended to employ a standard nomenclature throughout the book in order to avoid confusion and misunderstanding.

1. *Clamping Plate, injection side:* Forms the base of the fixed mould half; sits on the fixed platen of the machine and is clamped onto it either directly with screws or by means of clamps. Made out of MS or EN 8.

2. *Cavity Plate:* Houses either the outer form of the product directly or an insert containing that shape. Is fixed with the clamping plate (Pos. 1). Made of MS, EN 8, P 20, hardenable alloy steels and in special cases, out of light metal alloys.

3. *Core Plate:* Houses the core, Made of MS, EN 8, P20, hardenable alloy steels or light metal alloys.

4. *Cover Plate:* Supports the core from behind. Made of MS, EN 8, P20.

5. *Distance Blocks:* Maintain distance between the cover plate and the clamping plate (Pos. 8) on the ejection side, thus creating room for the ejection unit and its movement. Made of MS, EN 8.

6. *Ejector Holder Plate:* Holds the ejectors. Made of MS, EN 8.

7. *Ejector Cover Plate:* Supports the ejectors. Made of MS, EN 8.

8. *Clamping plate, ejection side:* Forms the base of the moving mould half and holds components 3, 4, 5 together; sits on the moving platen of the machine and is clamped onto it like pos. 1. Made of MS, EN 8.

9. *Register Ring:* Centres the mould half on machine platen. Fixed with the clamping plate. Made of MS, EN 8.

10. *Sprue Bush:* Acts as the seat of the machine nozzle and as the first point of entry of the melt in the mould. Made of hardenable alloy steel, EN 24.

11. *Sprue Puller Bush:* Pulls out the sprue with the help of reverse taper/undercut. Made of hardenable alloy steel, EN 24.

12. *Cavity Insert:* Houses the outer contour of the article. Made of hardenable, non-shrinking alloy steels.

13. *Core Insert:* Forms the inside shape of the article. Made of hardenable, non-shrinking alloy steels.

14. *Guide Pillar:* Aligns the mould halves accurately through a matching bush.

15. *Guide bush:* Aligns the mould halves accurately through the guide pillar. Made of hardenable steels.

16. *Cylindrical Dowel:* Aligns the mould plates upon assembly. Made of hardenable steels.

17. *Ejector Rod:* A link between the ejector plates in the mould and the ejection system of the moulding machine. Made of hardenable steels like EN 24.

18. *Guide Bush:* Guides the ejector plates. Made of hardenable steel.

19. *Return Pin:* Pushes the ejector plates back to their initial position. Made of nitride hardening alloy steels.

20. *Ejector Pin:* Pushes out the moulding from the core or cavity; made of nitride hardening alloy steels.

21. *Sprue ejector pin:* Pushes out sprue and runner. Made of nitride hardening alloy steel.

22. *Transport Strip:* Holds both mould halves together during transport, loading and unloading as well as storage. Also used for lifting the mould with an eyebolt. Made of MS, EN 8.

Note: The guiding components 14 + 15, having a H7-g6 fit, are bound to have some clearance which may not be acceptable in exceptional cases of extremely precise articles. In such cases, these are supplemented with an additional set of guiding members consisting of a conical pin and a corresponding bush as shown in Fig. 4.2.

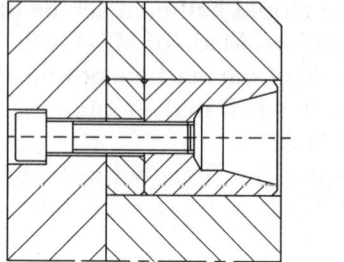 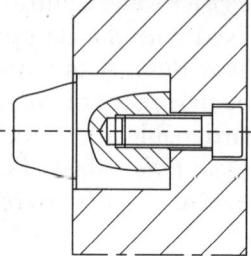

Fig. 4.2

a. Depicts the feeding system comprising sprue, runners and gates.
b. Shows the cavity or the shape of the article to be moulded.
c. Represents the cooling network for circulation of cooling medium.
d. Marks the ejection system and the device to move them.
e. Denotes the parting line where the two mould halves meet and part.

Understandably, the mould must open at a suitable place to enable extraction of the article out of the cavity. In other words, it must be made of at least two units, which are held together during injection but go apart for demoulding. The plane of meeting is called mould parting line. Depending upon the shape of the product, the cavity may be housed in one or both mould halves.

The mould, depicted here, is of a so-called 2 plate type, because it consists of two units or two halves, viz. the injection or the fixed half and the ejection or the moving half. As the name indicates, the fixed half is attached to the fixed platen of the injection moulding machine and the moving half to the moving platen.

Although there are many other types of moulds, which have special components, the ones shown here occur in almost all variations. However, the common features of all injection moulds are injection (feeding), core and cavity, parting, cooling and ejection.

Almost all mould components except the core and cavity inserts are available as standard parts in different sizes and materials.

The Feeding System

The feeding system in the mould is the path of the melt from the source, that is the nozzle of the injection moulding machine, to its final destination, viz. the mould cavity. It should perform this function with as little loss in material, temperature and pressure as possible. It is also required to distribute the melt equally to all cavities of a multi-cavity mould and fill them simultaneously with same pressure and temperature. Another condition of equal importance is that the feed paths remain open till the end of the holding phase but should be adequately solidified to be ejectable when the article is demoulded.

The feeding network has three distinct sections. They are:

The Sprue (Fig. 4.3)

It is the straight path of the melt from the machine nozzle to the primary runner or the melt distribution network. It is round in cross-section and has a taper along the length for easy removal. It is usually housed in a round standard part called sprue bush, which has a suitable spherical depression on the receiving end as seat for the machine nozzle. The starting diameter of the sprue is about 0.5–1 mm larger than the bore of the nozzle to accommodate any misalignment. The sprue is provided a taper of at least one degree for easy removal.

The sprue ends in the main distributing channel or the primary runner but it may also directly lead to the cavity in the case of a single cavity mould.

Another variation of this is an inclined sprue (Fig. 4.4). This design is usually resorted to with single cavity moulds when the cavity is situated in the centre of the mould but the sprue is not

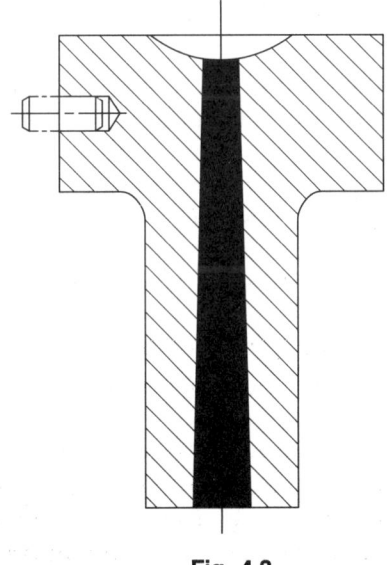

Fig. 4.3

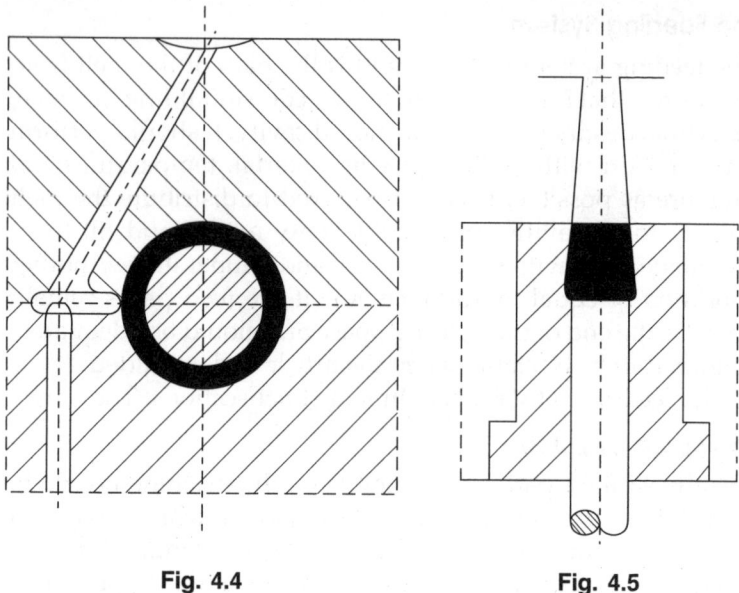

Fig. 4.4   **Fig. 4.5**

permitted to be placed right on the product. The angle of inclination from the mould axis should not exceed 30 degrees.

The Cold Slug Well

The melt in the tip of the nozzle loses heat to the cold mould when resting on it. A cold slug well is provided in the mould at the end of the sprue to receive and retain the colder melt, preventing it from flowing to the cavities (Fig. 4.5). The cold slug well with reverse taper also serves as a sprue puller. An ejector is positioned under the well to knock it out.

The extent of the taper depends upon the rigidity of the plastics. Whereas a reverse cone of 5–8° can safely pull out a sprue of polystyrene, 10–15° may be needed for softer materials like polyethylene. Too big a taper may pose problems in ejection. The depth of the well equals the smaller diameter of the cone.

An alternative design, combining the two features, viz. the cold slug well and the sprue puller, is shown in Fig. 4.6. It is as effective as the foregoing design, is simpler and cheaper but the sprue may not fall off automatically after ejection.

The Runner

It is the linking channel, which carries the melt from the end of the sprue to the edge of the cavity. The main or the primary runner may have secondary runners, smaller in size, which may be needed to lead the material from the primary runner to the cavities, depending upon the number and arrangement of the cavities.

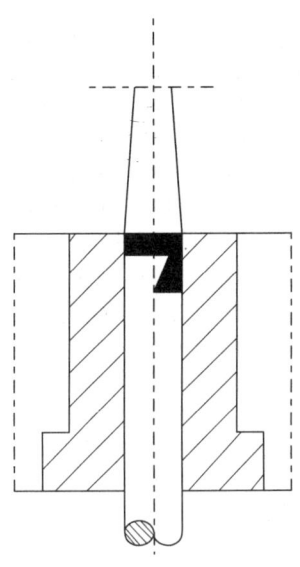

Fig. 4.6

A good runner should convey the melt with minimum loss of heat. In other words, the most effective runner would have maximum cross section with minimum circumference. It may be round, square, rectangular or semi circular in cross section (Fig. 4.7). For example:

Circle,10 mm. dia.   Cross section =78.54 mm² Circumference=31.4 mm.
Square, 8.86x8.86   Cross section=78.54 mm² Circumference=35.4 mm.
Rectangle, 12.5x6.25   Cross section=78.54 mm² Circumference=37.5 mm.
Semi circle, R=7.07   Cross section=78.54 mm² Circumference=58.56mm.

The round type is the most efficient one as it offers maximum sectional area for flow with minimum circumference through which the heat is dissipated to the mould. The same area of cross section in other forms mentioned above will have larger circumference. The round runner, however, has to be cut in the cavity as well as the core plates whereas square, rectangular and semi circular runners can be machined on one side. A modified trapezoidal runner with a semi circular base, placed

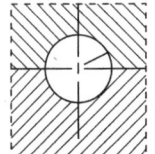 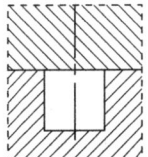 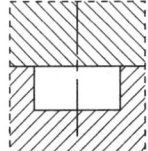 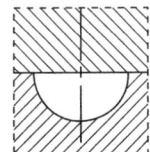

Fig. 4.7

in one plate, is the next best solution (Fig. 4.8). A total draft of 10–20 degrees facilitates positive ejection.

The runner or the secondary runner ends about a millimeter before the cavity.

Runner Size: A runner should bring the melt to the cavity with as little loss of temperature and pressure as possible and should freeze after the gate has closed. As mentioned earlier, the length of the runner is directly dependent upon the number of

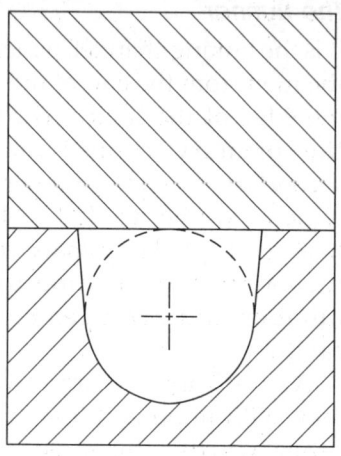

Fig. 4.8

cavities and their layout in the mould. The size of the runner, viz. its cross-section depends upon the particular plastics and its viscosity, the volume of the moulding, the distance of the cavity from the source of the melt etc. It is very difficult to provide exact data in the face of so many variables. A general rule, applicable universally, is that the runner should be at least as thick at the point of gating as the wall of the article there. In case of multi-cavity moulds, it is generally the secondary runner which feeds an individual cavity.

Though an oversized runner ensures supply of the melt beyond the point when the gate freezes, it is accompanied by disadvantages of material wastage and cooling time prolongation.

The size of the secondary runner should be chosen to maintain a uniform speed of flow. For example, if the main runner splits into two secondary branches, the total cross-sectional area of the later should equal that of the former.

Though desirable, it is sometimes not practicable to place all cavities equidistant from the sprue. Sometimes the cavities can only be arranged in a row and have to be fed one after the other. Simultaneous filling in such cases is achieved by manipulation of individual gates through trials. The cavities closer to the source of the melt are provided with smaller gates and those placed farther with bigger ones. It must, however, be underlined here that the balancing achieved is valid only for that particular grade of thermoplastics. The filling pattern changes as soon as another material grade with different flow characteristics is employed. Similar results may also be obtained by varying the size of the primary runner in different sections or by increasing the size of the secondary runners as the melt travels away from the central source. A similar problem of uneven filling is encountered in case of the family moulds (moulds housing cavities for dissimilar articles). Because of this reason, fabrication of family moulds should be discouraged.

The Gate

It is the ultimate entrance of the melt into the cavity from the end of the runner. The one shown in the introductory diagram is a side gate, also called edge gate.

A side gate needs to be trimmed off, leaving a small mark on the product.

The introductory mould design (Fig. 4.1) shows these features, viz. the sprue, the cold slug well and the runner with an edge or side gate. The other common types of a side gate are the submarine gate, the fan gate, the film gate, the tab gate, etc. which are elaborated below.

The Edge Gate

In multicavity moulds, runners act as distributors of the melt and bring it close to each cavity. A thinner channel forms the connection between the two. It is called an edge gate. The most efficient cross section for a gate is a circle which can be used if the cavity is placed in both mould halves. It needs to be machined on both sides. The next best alternative is a trapezoidal gate like the runner shown in Fig. 4.8. However, gate with a rectangular cross section, placed in one mould half, is easy to machine and easy to modify during the try-out. A gate may be as deep as two thirds of the article thickness at the point of entry (for very thin parts, it may equal the article thickness). The breadth is generally twice the depth. The length of the gate, also called the land, is 1.0 mm or preferably less.

It remains connected with the moulding and is severed after ejection, leaving a mark. The edge gate is easy to make and easy to enlarge during trials.

Edge gating has been shown in the introductory mould design (Fig. 4.1).

The Submarine Gate

It is a conical tunnel, going from the end of the runner to the wall of the cavity at an angle through the solid insert. It is round in cross section and has its smallest diameter where it enters the cavity. The angle of inclination with the mould parting line should be at least 45 degrees. In principle, it is a side gate but has the advantage that it gets sheared off automatically, leaving an elliptical mark on the wall, when the mould opens or during ejection. It is practicable with all thermoplastics (Fig. 4.9).

Another version of the submarine gate is shown in Fig. 4.10. The gate is formed like the frustum of a cone, protruding partly in the cavity. The smaller diameter of the frustum is 2.5–3.0 mm. It has the advantage of containing a fluid "soul" or core. It leaves a crescent shaped mark on the product. The angle of its inclination with the mould axis should be same as in the previous version.

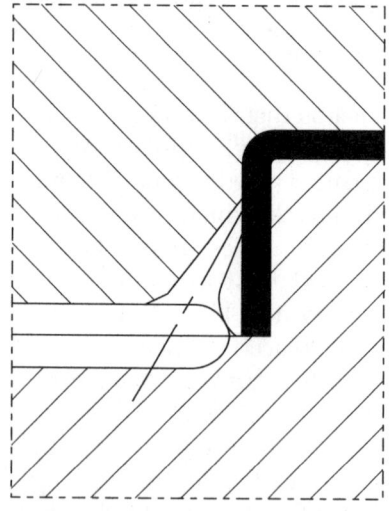

Fig. 4.9

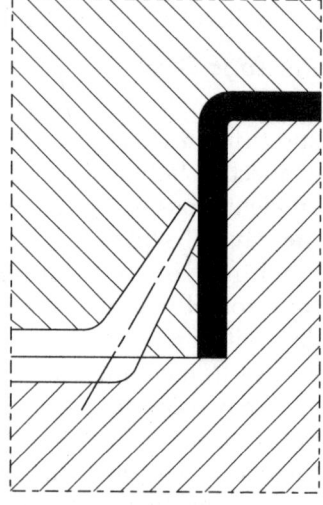

Fig. 4.10

A gate mark on the visible surface of some sophisticated product may not be acceptable. In such cases, the submarine gate may be placed on an auxiliary detail, such as a thin leg (Fig. 4.11), at an invisible spot. The leg can be removed after ejection.

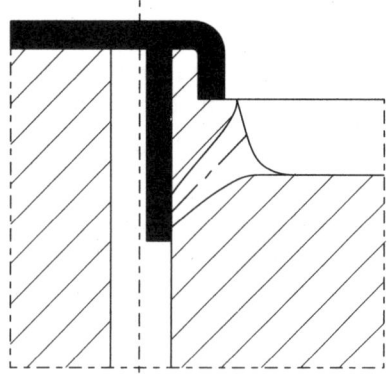

The Banana Gate

It is a further refinement of the submarine gate and has

Fig. 4.11

the form of a semi circular tunnel, starting from the end of the runner and ending on the underside of the moulding. It is resorted to when a mark on the visible face of the article is not permitted. It can be used with almost all materials except the very brittle ones (Fig. 4.12). As shown, it has to be housed in a split insert for ease of manufacturing. However, standardised solid inserts are also available now with various gate sizes.

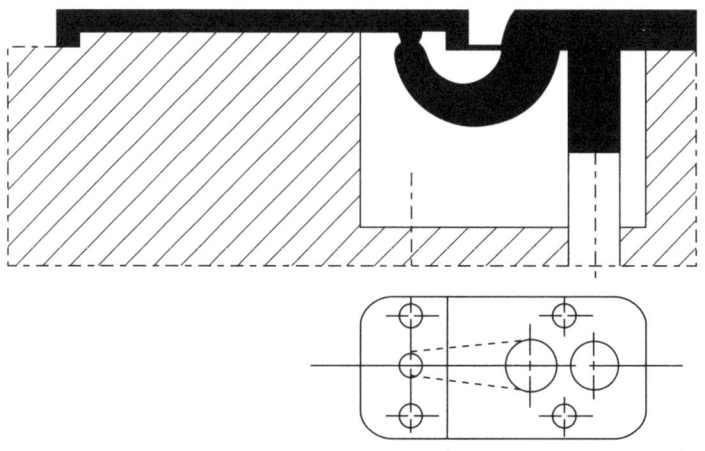

Fig. 4.12

The Film Gate

It is employed for large, flat articles. Spread over the complete side of the cavity, the film gate ensures an even filling, causing less warpage and leaving no flow marks, but has to be machined off after ejection (Fig. 4.13).

The Fan Gate

It serves the same purpose as the film gate. The difference lies in the configuration, which gives it the appearance of a fan, emanating from the end of the runner and encompassing a part of the edge of the cavity (Fig. 4.14). The breadth of the gate is computed by dividing the cross-sectional area of the runner by the depth of the gate.

The Tab Gate

It is an indirect form of gating as it connects the runner with a tab on the moulding. It is employed for some specific materials like acrylics, polycarbonate, acrylonitrile styrene (SAN) but also in cases where flow marks are difficult to eliminate. The material is first introduced into a tab from which it flows evenly into the cavity. The tab, which may be as deep as one half or

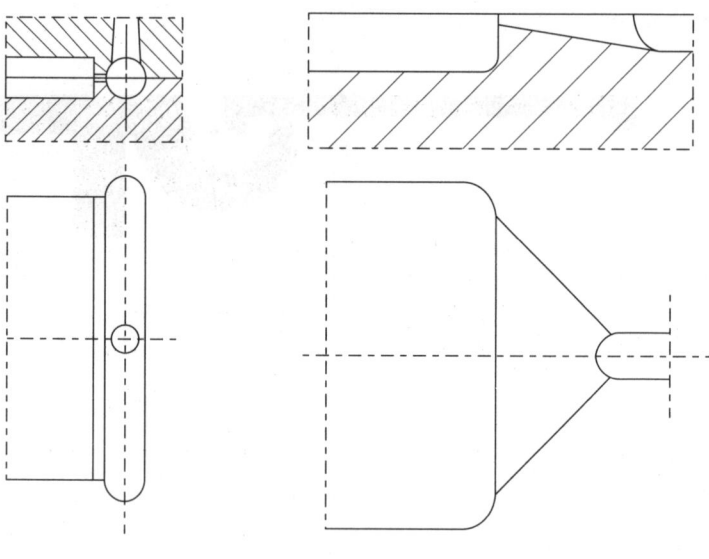

Fig. 4.13   Fig. 4.14

even two thirds of the wall thickness of the moulding, has to be machined off (Fig. 4.15). The breadth of the tab may be 6 mm or more.

The Ring Gate (Fig. 4.16)

It is a film gate for round articles. The runner is placed like a round ring outside/ inside the article and a thin film gate connects both all along. It ensures uniform filling and roundness of the product but necessitates after-operation for removal. In uncritical cases, where some deviation in roundness is permitted, three or more gates may be used instead

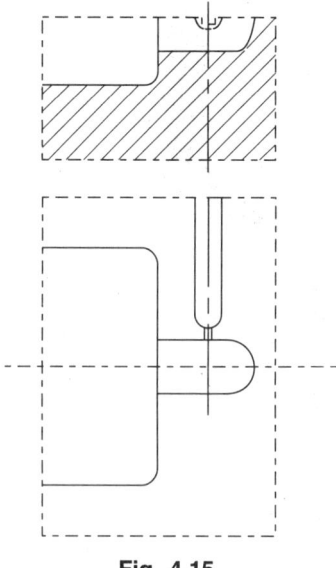

Fig. 4.15

of a continuous film. The degating is more convenient. It may also be possible to use self shearing submarine gates.

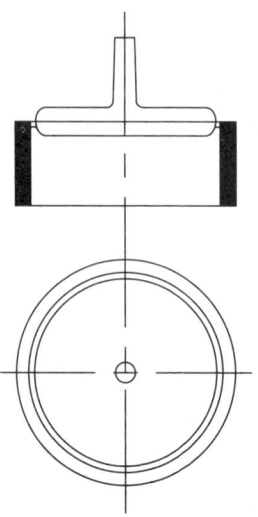

Fig. 4.16

The umbrella gate: It is like a conical ring gate. The runner is connected to the round article through a film gate all along the rim and fills it uniformly ensuring roundness. It can also be modified to feed at three or more places. That facilitates centring of the core (Fig. 4.17).

The Sprue as Gate

Single cavity moulds have usually the product cavity placed in the centre of the mould. Here the sprue can take over the function of the runner as well as that of the gate (Fig. 4.18). The sprue is ejected along with the moulding and removed in an after-operation, leaving a round mark which may not be found

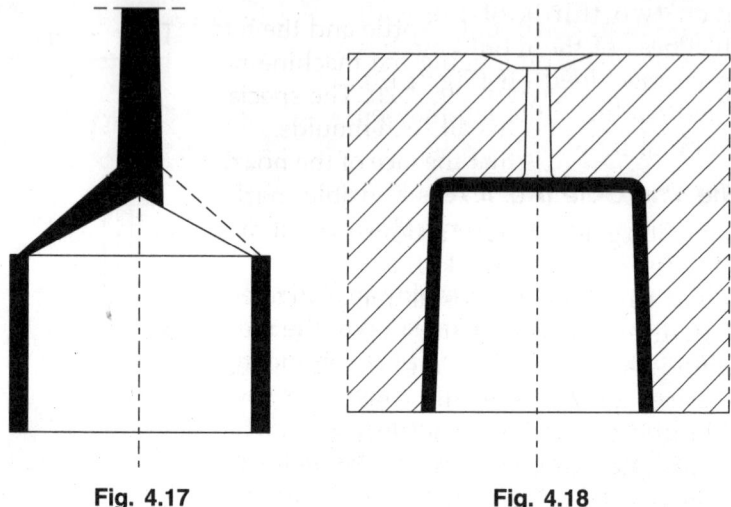

Fig. 4.17　　　　　　　　**Fig. 4.18**

disturbing if it is on a invisible spot, for example, on the back of a bucket or a box.

The Pin Point Gate

The sprue gate can be converted into a self severing pin point gate with the help of a special sprue bush (Fig. 4.19). As the injection unit of the moulding machine recedes after injection and the holding period, a part of the sprue bush is pushed back pneumatically, severing the gate and ejecting the sprue backwards. The gate is circular in cross section, 1–1.5 mm. in diameter and in length and tapered backwards so that it breaks at the cavity, leaving a very small vestige. The included angle of the conical gate may be around 30°. The opening in the cavity is countersunk to a depth of 0.5 mm to provide a defined breaking point.

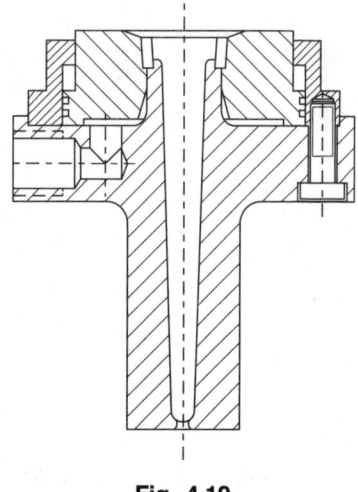

Fig. 4.19

A pin point gate without a sprue and the moving bush can also be realised by prolonging the machine nozzle up to the mould cavity as shown in Fig. 4.20. The special nozzle can be standardised so that it fits all such moulds.

The disadvantage is that the face of the nozzle, which is at a very high temperature, leaves a visible mark with different surface finish on the moulding. However, it may not be visible in some cases.

Another alternative is to employ an elongated, self insulating machine nozzle, reaching almost up to the mould cavity, with its tip fashioned out of a good heat conducting metal such as copper or preferably beryllium copper. As shown in Fig. 4.21, the mould contains a special sprue bush in which the machine nozzle sits with a gap all around. The melt from the first shot creates an insulating sheath of plastics around the tip of the nozzle and prevents loss of heat so that the nozzle remains hot in a cooled mould and the product surface does not show any blemishes caused by high temperature. The sheath is thinner in front (about 0.2–0.3 mm.) but gets thicker gradually around the shaft. The thin film in front is thick enough to be effective as insulation but not too thick to block the next injection. The gate length in the sprue bush is about 1.5 mm. The heat loss

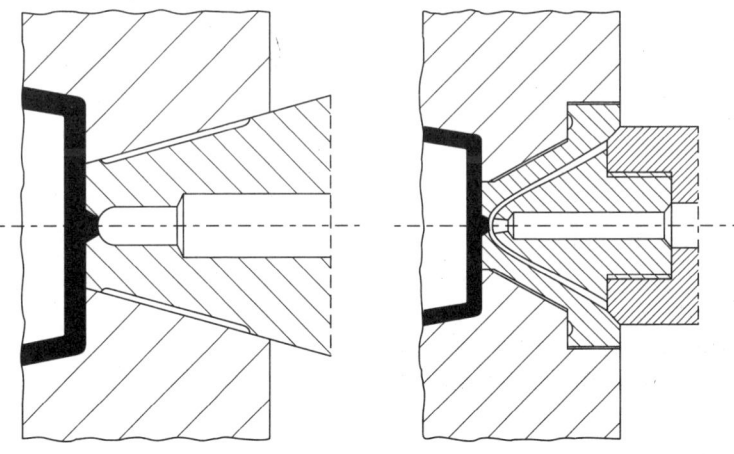

Fig. 4.20   **Fig. 4.21**

through the sprue bush is kept to the minimum by reducing the area of contact between the mould and the bush.

Three Plate Mould

The side gate in different forms as discussed above, is simple in mould configuration and economical in cost. The mould is required to open only at one level to eject runners, gates and the mouldings. It is called a two plate mould. The side gate, however, is primarily suitable for shallow articles with uniform wall thickness. It leads to blemishes such as air entrapment, short shots and burn marks in case of tall parts like boxes, tubes, tumblers, etc. The material injected at the edge of the rim travels around the rim as well as upwards. If the flow path along the rim is shorter than that across the top, material will fill first around the rim and seal the escape path of the air contained in the mould before complete injection (Fig. 4.22). The melt stream, travelling towards the top, goes on compressing the entrapped air till it gets so overheated that it burns the plastics around it. The problem would not arise if the product could be injected in the centre. The melt advances like a uniform front, driving the air before it towards the mould parting line, where it can escape easily. Such an arrange-ment is possible in a single cavity mould, where the sprue can function as a direct gate (see Fig. 4.18). The sprue will have to be removed in an after-operation and a big mark will be visible. Another alternative is a pin point gate by means of special sprue bush and nozzles as shown in Figs 4.19 to 4.21.

To feed more than one such cavities centrally (or from above) with pin point gate calls for bifurcation of the runner at a level other than the parting line (Fig. 4.23). In order to remove the runners, the mould

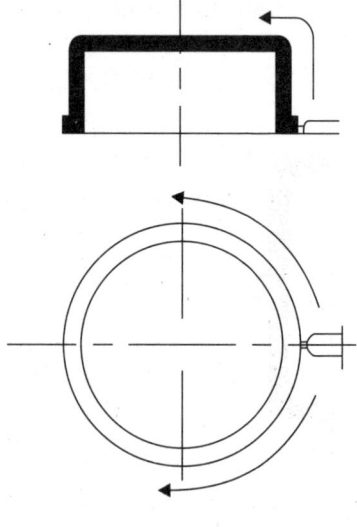

Fig. 4.22

must be opened at their levels also. This type of mould would consists of three units and is named as a "three plate mould". It will be discussed in detail in the section 4.2 "Special Mould Designs".

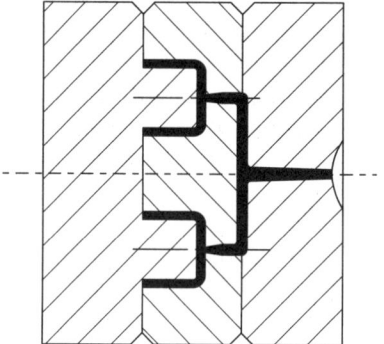

Fig. 4.23

The Self-Insulating Runner

The advantage offered by the three-plate system, viz. the freedom of location of the gate, is invaluable. However, the disadvantages of the runner network at a different level and the complications caused by an additional opening and the arrangement to eject the runners have encouraged the enterprising designers to find ways and means to circumvent the drawbacks while retaining the positive points of the system. The bad heat conductivity (that is, good thermal insulation) of plastics led to the experiment of making the runner cross-section sufficiently large so that the frozen outer layer insulated the melt in the centre and kept it fluid till the next shot. It worked, especially with thermoplastics with a broad melt temperature range. This was the birth of the runnerless moulding. This system has been named as the self-insulating runner. The gate is of the pin point type.

As mentioned before, plastics are bad conductors of heat. If the runner were sufficiently thick, the plastics in the middle of it, insulated by the frozen outer layer, would remain hot and molten for a long time (Fig. 4.24). Successive shots could be pumped through this

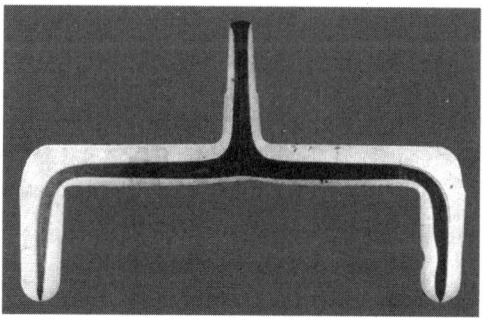

Fig. 4.24

insulated "pipe", without taking out the runner. The mould would behave like its two-plate counterpart and also have advantages of a three plate mould.

The system is simple and cheap and works very well for some thermoplastics if the cycle is short and regular (see Special Mould Designs, 4.2 for design details).

The Hot Runner

The foregoing solution of runnerless moulding suffers from severe limitations. It is not applicable universally. Any disturbance or interruption in cycle leads to freezing of the runner necessitating its removal and starting the operation afresh. The hot runner concept aims at keeping the runners hot at desired temperature with positive means independent from the cycle. It can be employed for the moulding of almost all thermoplastics.

Hot runner system allows freedom of location of the gate. Standard systems are available in diverse variations for different materials, products and shot weights. Some designs have been described in detail in section 4.2 on Special Mould Designs.

The Stack Mould

A stack mould may be regarded as amalgamation of the three plate and the hot runner systems. It is basically a three plate construction but instead of the runner network, another set of cavities is accommodated at that level. The feeding system is mostly a special hot runner housed between the two levels, which feeds both sets of cavities.

A stack mould virtually comprises two 2-plate moulds, placed back to back, thus having twice the number of cavities in aggregate. It can almost double the production from the same moulding machine. For opening on two level like a three plate mould, the stack mould also needs special devices. The system is described in detail in section 4.2.

In principle, a stack mould can have more than two stacks.

Number of Gates

It may not be possible to fill the cavity with a single gate if the article is very large. The ratio between the length of flow and the wall thickness for the polymer in question is a good indicator for deciding the number of gates. Another consideration, no less important while deciding the number of gates, is that the pressure needed for filling should not be more than the half of the machine capacity. Overpacking results in warpage.

More than one gate may also be called for to ensure balanced filling of the cavity. For example, melt from a centrally placed gate on a rectangular product will cover the shorter distance first and reach shorter sides of the rectangle later. During this period, the melt which has already touched the cavity wall will be under pressure and will get overpacked. It is bound to lead to internal stresses which cause warpage. Two gates, located to provide equal flow lengths, will eliminate the problem (Fig. 4.25).

More gates, however, give rise to weld lines and may cause air entrapment in some cases. The quality of weld lines, i.e. the strength and appearance of the joint, depends upon the temperature of the meeting streams and their pressure.

A gate is also the point of maximum internal stress. More gates also mean more centres of stress concentration. This results in uneven shrinkage and warpage in extreme cases.

Guidelines for location of the gate. The function of a gate is to fill the cavity in a manner that ensures a sound moulding, free of faults. Its location has a far reaching influence on the quality of the product. Following guidelines deserve due consideration while determining the gate location.

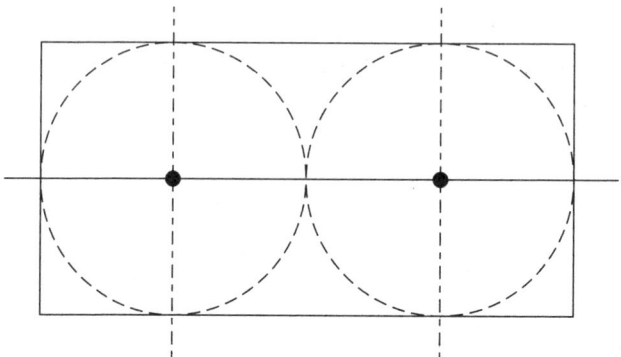

Fig. 4.25

1. The gate should be so located that the length of flow is same to all end points.
2. The gate location should ensure rapid and uniform filling.
3. It should ensure minimum runner length.
4. It should ensure melt progress as an advancing front.
5. It should facilitate air expulsion and prevent air entrapment.
6. It should direct the melt stream towards a restriction and prevent jetting.
7. It should be close to the thickest section for the holding pressure to be effective longer.
8. It should banish the unavoidable weld lines to harmless areas.
9. It should not cause deflection in slender mould cores.
10. It should be away from the load bearing section as the gate area has maximum stresses.
11. It should not make the mould unnecessarily more complicated.
12. The gate location should not interfere with the function and appearance of the article.

Besides location, the size and the shape, too play an important role. For a successful moulding operation, the gate should:
- Leave a small vestige
- Prevent incomplete filling
- Prevent premature freezing
- Prevent jetting
- Permit unrestricted flow
- Provide correct shearing
- Hide/disperse cold slug
- Prevent drooling and stringing
- Prevent blushing
- Prevent degradation
- Prevent warping

Arrangement of Cavities

In a multi-cavity mould, the melt entering the mould at one central point, viz. the sprue, travels to individual cavities through the runners and gates. If their lengths and sizes are not identical, the cavities would not fill at the same time. When the holding pressure sets in, the cavities which filled first would get overpacked and may tend to flashing while those which filled later would be underpacked and may have sink marks and voids. The difference in the pressure and its duration leads to difference in dimensions as well as residual stresses. Such feeding system is termed as unbalanced. The safest method of a balanced filling is to place the cavities equidistant from the sprue and have a separate runner to each cavity. This may prove impractical as it necessitates their placement in a circle which would become bigger with more cavities, making runners longer. The mass of runners may even overweigh that of the mouldings. The solution is to divide the cavities in a number of groups and feed the group by a main runner from which secondary runners lead to individual cavities. Figure 4.26 shows a few arrangements of balanced placement of cavities.

In some exceptional cases, it may not be feasible to locate the cavities in the ideal manner. Sometimes, the linear arrangement is unavoidable. The length of the runner to each cavity will be different. Here, the balancing is effected by

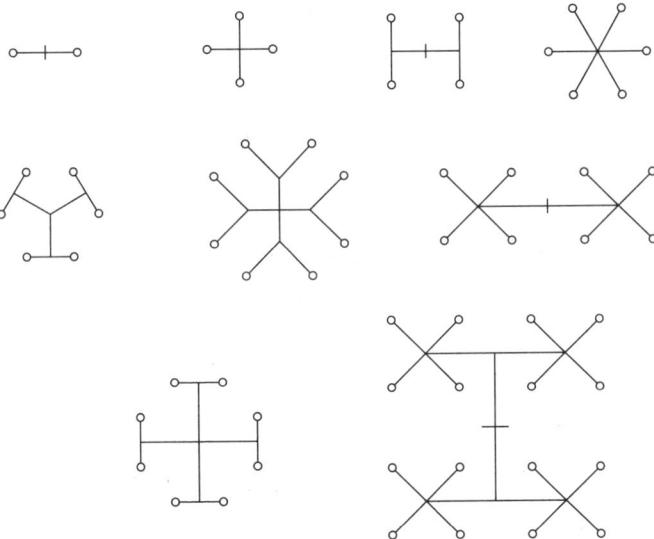

Fig. 4.26

making the gates different in size so that the filling takes place simultaneously. To start with, all gates are made equal and then enlarged step by step during the trial on the machine. It must, however, be borne in mind that the balancing achieved is valid only for that particular material. The flow behaviour will be different with a different grade.

Number of Cavities

The maximum number of cavities in a mould for a particular machine is directly dependent upon the locking force of the machine, the projected area of the article and the effective injection pressure. The last factor is not identical with the set pressure or the one displayed on the pressure gauge. The melt loses some pressure till it reaches the end of the cavity. By the time a cavity is completely filled, a part of the melt has partially frozen and does not exercise full pressure. Effective pressure may vary between half to two thirds of the pressure displayed. It is influenced by the melt and mould temperatures, the length of travel and the speed of filling, among other factors.

The maximum possible number of cavities, however, may not always prove to be the most economical for a given quantity. Though the production rate increases and the moulding cost decreases with the increasing number of cavities, the cost of the mould also goes up.

The profitability of multi-cavity moulds is directly linked with the total quantity to be moulded. The most economical number of cavities can be computed from the following data:

Q = Total number of mouldings to be produced
M = Mould manufacturing cost per cavity in Rupees
T = Moulding cycle in seconds
η = Moulding efficiency in decimal
R = Moulding machine rate per hour in Rupees
N = Economical no. of cavities

$$N = \sqrt{(Q \times R \times T)/(M \times 3600 \times \eta)}$$

The number of cavities thus calculated may not be an integer. It is advisable to round it up or down to a whole number depending upon the value of the fraction. It requires further examination whether some unusual features of the article come in the way of accommodating the calculated number of cavities in the mould commensurate with the size of the platen and capacity of the machine. These features may be the shape of the product such as rings and frames with large circumference but relatively small projected area, or those with less area but large volume, exceeding the shot weight or products with undercuts requiring sliders on sides, etc. Another vital check is the comparison of the locking force of the machine with that required for the selected number of cavities.

Mould Venting

Before injection, mould cavities contain air which must make place for the plastics. The advancing melt stream pushes the air farther and farther and in most cases it can escape at the end of the path through the parting line where the two halves of the mould meet, unless the mating surfaces are very finely ground. However, the configuration of the article or the position of entry of the melt may make the melt seal the parting line before all air could escape as already discussed and illustrated in Fig. 4.22. The entrapment of air may also happen in article features like lugs, ribs and bosses. The advancing melt compresses the air till it also achieves a high pressure and hinders the flow or it gets so overheated that it singes and chars the polymer. The phenomenon is rightly termed as the diesel effect.

The normal escape route for the enclosed air is the parting line, which may be foreseen with additional venting channels on the face, not deeper than 0.04 mm., at crucial points where the melt reaches last. Ejectors and other moving mould components like slides and cores and split cores are other effective escape routes for the entrapped air. The joints of the unmoving cores and inserts also provide a path for the air but they are likely to get choked with time by the escaping volatile additives in the plastics and need occasional cleaning.

Additional ejectors at the points of entrapment of air serve as venting aids but in many cases it may not be possible to position ejectors there. A viable solution is to incorporate sintered metal inserts at the critical locations, providing holes at their back for the final escape of air. The sintered metal inserts too require occasional cleaning. They should be easy to remove and refit.

Air in the feeding channels of the cold runner moulds is driven by the melt into the cavities. A venting groove, as broad as the runner and about 0.03–0.05 mm. deep at the end of primary runner, would let the air out.

If the melt is parted in the mould by some design feature like a pin forming a hole, the streams form a weld line when they meet again. The joint would be weak if some air is trapped between the two fronts. Venting provision at that juncture can contribute to the strength of the weld.

Mould Cooling

Plastics is injected into the mould in molten condition at an elevated temperature. For most of the commodity plastics, it is around 200–250°C; for engineering platics the melt temperature is even higher. The mould gives the melt the intended shape of the end product but it also does something more than that. It acts as a heat exchanger. It cools down the hot, injected plastics mass to a temperature where the moulding becomes rigid and can be ejected out of the mould without distortion. To facilitate the dissipation of heat of the injected melt, moulds are provided with cooling channels for circulation of a cooling medium, usually water, but sometimes also oil at elevated temperatures—in case of some engineering plastics it may exceed 100°C. It is nevertheless referred to as cooling because the melt is cooled, even though the mould is being heated. While discussing the moulding technology, it was observed that the cooling time constituted the major part of the moulding cycle. Consequently, design of the cooling system in a mould deserves proper analysis and careful layout to achieve the most efficient heat transfer. A longer cooling time means a longer cycle, which in turn means less production and consequently higher production cost per unit.

Before designing the cooling layout of a mould, one must consider the following factors:

a. The shape of the moulding.
b. The pattern of cavity filling.

c. The hot spots.

d. The thin and thick sections or the zones of less or more heat.

e. Inaccessible hot areas.

Design of the Mould Cooling

Cooling would be most effective if it conforms to the shape of the product. In other words, for rectangular articles, cooling should have a rectangular layout and for round mouldings, it should be spiral or circular in design. Simultaneously, the cooling should be more intensive at the hot spots. It can be achieved by providing more channels near the thicker sections and by arranging the entry of the cooler water near that spot.

A fact worth taking into account while designing the cooling layout is that the mould core is enclosed by the hot melt and its heat can only be conducted away by the incorporated cooling. Especially the corners of a core become heat accumulation zones and the melt cools there last. The cavity block, on the other hand, has more space to incorporate cooling channels and has all its outer surface exposed to atmosphere. A good part of its heat is dissipated outwards by convection and radiation. This phenomenon is observed in the moulds for box shaped articles (Fig. 4.27). The corners of the core constitute hot spots and cool slower than the walls. The corners shrink more and pull in the sides which results in caving-in of walls of the box like article.

Cooling changes the density of the plastics material. The change is more pronounced in the semi-crystalline polymers than in the

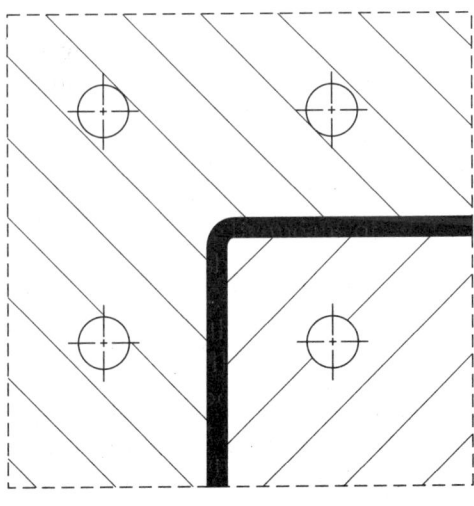

Fig. 4.27

amorphous ones. A non-uniform cooling of a semi-crystalline melt gives rise to different densities at different points, which in turn leads to warpage. Uniformity of cooling is, therefore, more important than its intensity.

The importance of uniform cooling is manifested distinctly in the form of warpage and rejection in products like compact discs and flat screens. In these cases, it is not only the temperature of the coolant but also its flow and the exactness of distances of the cooling lines from the cavity in both mould halves that influences the precision.

The side of the article with better cooling will solidify and shrink faster. The hotter side gets more time to cool and consequently shrinks more. The final product tends to bend with convexity on the colder side. It is, therefore, imperative that mould surface temperature be uniform on both sides of the part to prevent warpage. (The phenomenon may be made use of to counteract bending if one side of the mould cannot be cooled as efficiently as the other side. The opposite side may be kept hotter to equalize heat transfer and prevent differential shrinkage).

Types of Cooling

Cooling holes. It is the cheapest arrangement of cooling. Straight holes are drilled in the mould plates or inserts and interconnected to form a cooling network. This method of cooling is employed in case of rectangular cavities of low height. The rectangular cores can also be cooled with drilled cooling lines if the marks of the plugs on the inside of the moulding are acceptable. It is also resorted to in case of relatively cheap moulds irrespective of the shape of the moulding, where the price of the mould and not its efficiency are the deciding factors, for instance with experimental moulds and for those for small quantities. The usual hole diameters range from 6 to 12 mm. For large products, there should be more than one circuits, as otherwise the water may get too hot traversing a long path and mould may have widely different temperatures at the entry and exit points of the coolant (Fig. 4.28).

The simple formula of heat transmission highlights the factors which influence the quality and efficiency of cooling through drilled holes.

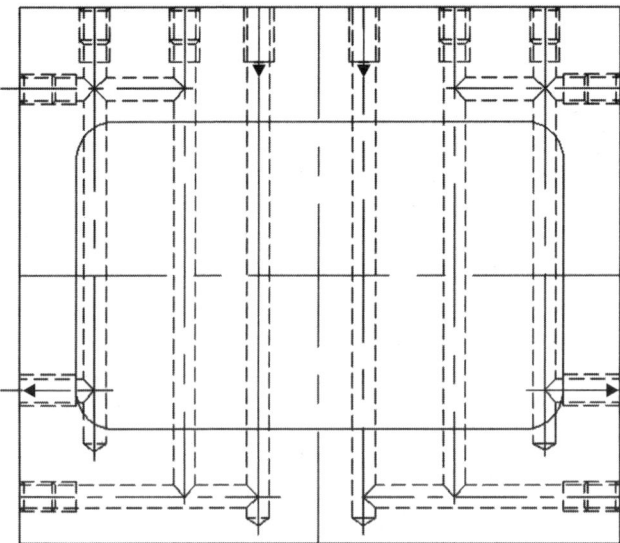

Fig. 4.28

$Q = \lambda.\Delta t.A/D$, where

Q   is the amount of heat transmitted

λ is the heat conductivity of the mould material

Δt is the temperature difference between the melt and the coolant

A   is the area of the cooling channel

D   is the distance between the cavity surface and the cooling channel

Obviously, more heat will be transferred:

- If the heat conductivity of the mould material is high (steel has relatively low heat conductivity)
- The temperature difference is greater (i.e. water is colder)
- The area of cooling channel is larger (i.e. the diameter of the cooling holes is bigger) or more channels are provided.
- The distance between the cavity surface and the source of cooling is shorter (i.e. the channels are placed closer to the cavity)

Steel is practically irreplaceable for mould components subjected to very high pressure and wear. Be-Cu may be regarded as an alternative. It has excellent heat conductivity which helps reduce the cooling time considerably but its lower tensile strength and hardness reduce the mould life.

It is an erroneous supposition that very cold water would expedite the cooling process. A thin layer of the melt solidifies on the cold walls of the cavity and acts as an insulating barrier between the hot melt behind and the source of cooling. Too cold a mould comes in the way of rapid filling and hinders crystallisation in case of semi crystalline polymers.

Cooling channels placed too close to the cavity surface result in temperature variations which will have adverse effect on the quality of the moulding. It will lead to different mould temperature at different points of the cavity (Fig. 4.29), which manifests itself as non-uniform shrinkage and distortion in the article.

Following guidelines (Fig. 4.30) concerning the size and distance of the drilled cooling lines ensure uniform and efficient cooling without impairing the strength of the cooled mould component.

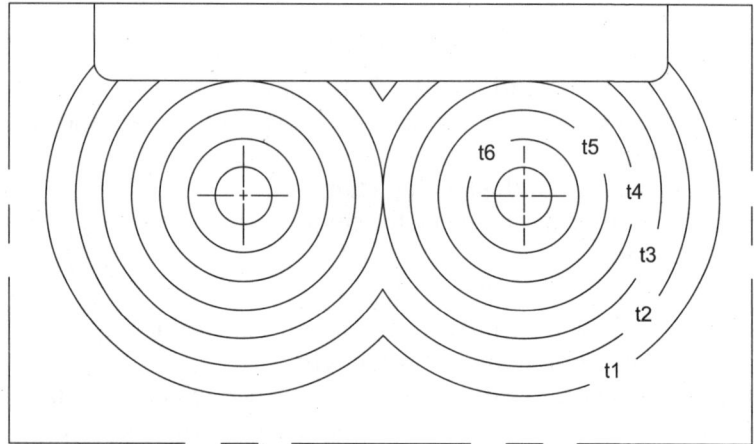

Fig. 4.29

d	=	dia of cooling hole	t = product thickness
X	=	2.5 d	Y = 2.5 d
d	=	6 mm for $t < 1$ mm	d = 8 mm for $t = 2$ mm
d	=	10 mm for $t = 3$ mm	d = 12 mm for $t = 4$ mm

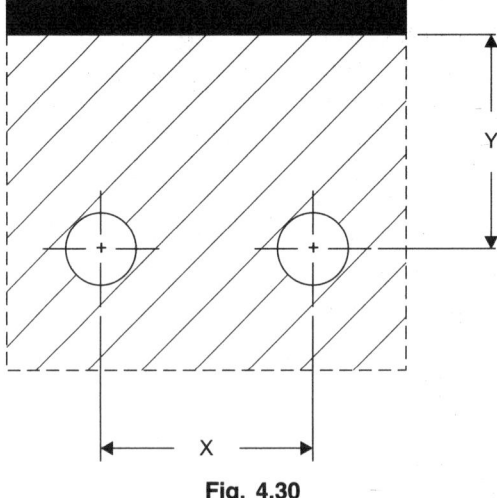

Fig. 4.30

Should it not be possible to accommodate adequate number of cooling holes because of some design features or ejectors in the way, it is better to opt for less holes placed at a distance larger than that recommended above. Though the cooling would be slower, it would be more uniform.

Long, slender drills are likely to stray from the intended line. Depending upon their diameter and depth, a clearance of 3–5 mm. must be foreseen between the cooling hole and any other perforations such as screw holes, ejectors, etc. in its path.

The rough surface of drilled cooling holes tends to gather deposits from water which reduce the cooling efficiency gradually. Water also causes rusting and the sharp grooves left from drilling form the starting point of cracks. It is therefore advantageous in the long run to ream the drilled holes and to coat them internally with chemical nickel.

The cooling efficiency can be enhanced significantly by introducing turbulence in the water stream. It can be effected by inserting twisted thin strips, preferably made of stainless steel, in the cooling holes.

Spiral Cooling

Cavities of round products like shallow boxes, plates or lids can be cooled more effectively if the cooling channels are cut in the form of a spiral at the back of the cavity plate and the inlet of the coolant is placed in the middle, close to the gate where the temperature is highest. It is cheaper and equally effective to machine concentric channels and form a continuous spiral like cooling path by blocking and joining them as shown in Fig. 4.31.

Bubbler Cooling

This type of cooling arrangement is usually employed to cool the cores of long cylindrical products such as tumblers, tubes, syringes, cones, bobbins and long slender containers (Fig. 4.32). The core is provided with a hole from the lower end and a pipe is fixed in the centre of the hole in such a way that there is clearance all around between the hole and the pipe. The cross sectional area of the clearance should be preferably less than that of the inlet hole. Cold water is introduced through the pipe to the hottest area. It returns around the pipe, cooling the walls of the hole and is led to the outlet. The bubbler arrangement is a balanced cooling method as the cold water is first led to the hottest spot, viz. opposite the gate. As the melt flows down, it

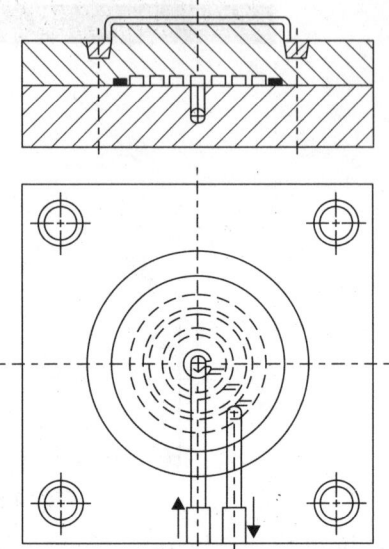

Fig. 4.31

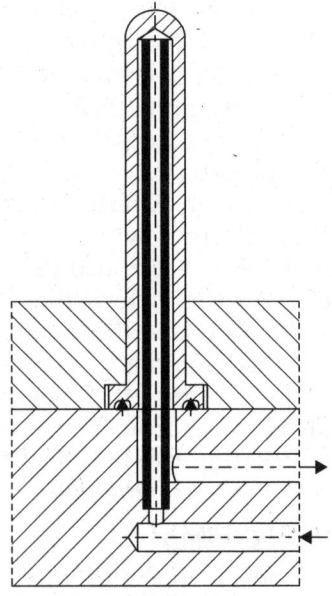

Fig. 4.32

gets colder and the cooling water gets warmer while bubbling down. This system can also be employed in case of deep and narrow articles of all shapes. Here, a number of bubbler holes are drilled in a row and depending upon their number, either provided with individual inlets and outlets or interconnected.

In most cases, the holes have to be interconnected as the cooling batteries of the injection moulding machines have a limited number of water circuits. There are two ways to interconnect the successive bubblers:

Parallel Connection

A common water inlet channel feeds all bubblers simultaneously and the used water is discharged in a common exhaust channel (Fig. 4.33). The main advantage of this arrangement is that all mould cores are treated alike as far as the temperature of water is concerned. However, even if one or more bubblers are partially or totally blocked, water flows in and out and it is not possible to detect the fault. Obviously, the cavities with the blocked water path would not be cooled as good as the others which fact may lead to variation in quality and even to rejection.

The system works well if water is soft and free of impurities and the pressure is adequate.

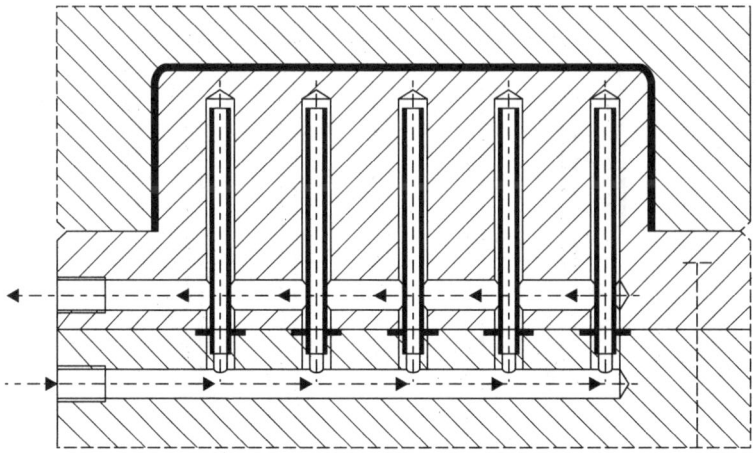

Fig. 4.33

Series Connection

In this design, fresh water is introduced in one bubbler and after circulating through it, it enters the next bubbler. In other words, exit of the first bubbler is the entrance of the next one (Fig. 4.34). Logically, there is bound to be some difference in temperatures between the successive bubblers. However, if any of the channels is blocked, no water would come out at the exit and the fault would be detected readily.

The difference of temperatures remains within acceptable limits if the water pressure is adequate and not too many bubblers are connected in a row.

Spiral Cooling of the Core

For large, round or rectangular articles such as mugs, buckets, radio and TV-housings, large box shaped articles, etc., insertion of a cooling core inside the mould core facilitates incorporation of uniform and effective cooling. For this purpose, a blind cavity is machined in the mould core from the backside and a cooling core, preferably made of a light metal like aluminium, brass or bronze, etc., is fitted therein. The cooling core is provided with cooling channels in the form of spirals, starting at the top where the mould temperature is maximum (Fig. 4.35). Water inlet and outlet are arranged in such a way that the coldest water enters

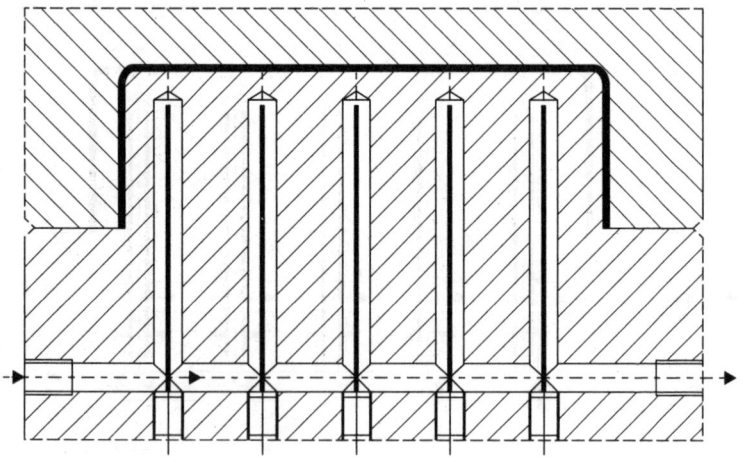

Fig. 4.34

close to the hottest area and travels downwards through the spiral, cooling the side walls. This is a balanced system of cooling like the bubbler arrangement. The principle can be applied in many variations depending upon the shape and size of the article. The cooling spiral can be broken into more than one independent segments with separate inlets and outlets for large articles.

The cooling core for a round article can be turned out of a light metal with a spiral going up to the top. A hole in the centre serves as water inlet. A milled channel connects it to the spiral so that water goes out cooling the core along its length. The inlet and exhaust may be designed as shown in Fig. 4.36.

The cooling core for a round article may be foreseen with a double spiral like a two start thread if the central axis is occupied by an ejector. Now one spiral serves as path for the water to travel up and the second spiral, which is connected here with the first one, forms the return path.

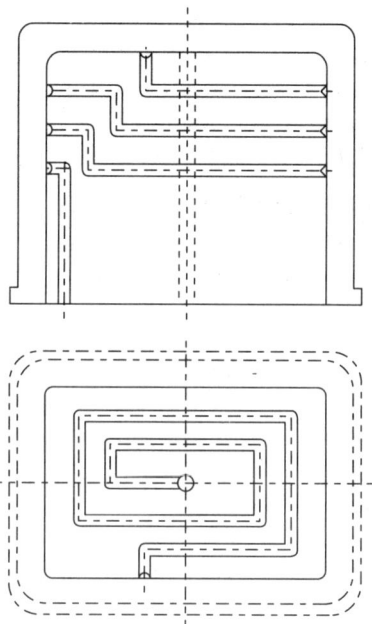

Fig. 4.35

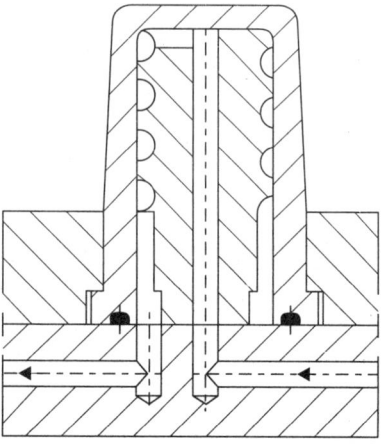

Fig. 4.36

Round cooling cores with a single or double spiral, machined out of a light metal, are available as standard components (Fig. 4.37).

Cooling Pipes

Cooling pipes or heat pipes represent the latest development in the mould cooling technology. They are employed to cool very slender, long cores (like those of the mould for an

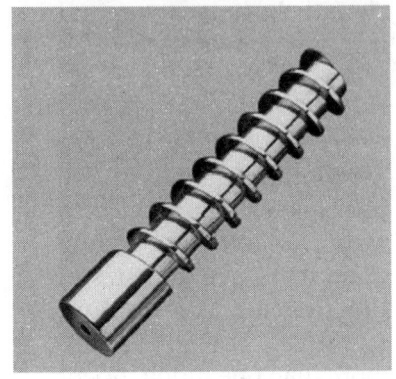

Fig. 4.37

insulin syringe), where it is difficult to use the bubbler system due to lack of space. The heat pipe is a hollow metal tube sealed at both ends and partially filled with a liquid which turns into a gas when heated. The heat pipe is snugly fitted in the mould part to be cooled with its lower end immersed in flowing cold

water. The fluid rises up to the hottest section through a fabric, gets heated, turns into a gas and then travels down to the lower end of the pipe that is being cooled constantly by a water circuit. The gas gives up its heat here and travels up again as a fluid (Fig. 4.38).

Heat pipes are available as thin as 3 mm. in diameter.

Copper wires and copper rods can also be employed in the same way as the heat pipes to conduct away heat from inaccessible mould sections though their efficiency would be much less. It can be improved by increasing length of the free end being washed by water.

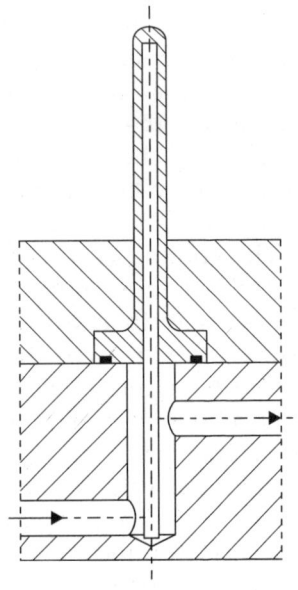

Fig. 4.38

Jacket Cooling of the Cavity: For tall, cylindrical products, the cooling of the cavity is equally important. The cooling layout for the cavity should also be executed in a balanced manner, with cold water entering near the hottest zone and proceeding in the same direction as the melt travels in the cavity. Figure 4.39 illustrates the design of the jacket cooling for a cylindrical component. A spiral is machined outside on the outer surface of the cavity insert and the water inlet is placed near the entrance of the melt. The outlet for the warmed up water is close to the rim of the moulding where the melt is colder.

A spiral may not be feasible for articles of short height. It is replaced by a single groove (Fig. 4.40).

Sealing rings, suitable for the temperature of the mould, must be provided to prevent leakage of the cooling medium.

The Flow of the Coolant

The most common coolant used is water which is not a good conductor of heat. If it is allowed to have a laminar flow, the heat from the melt will be conducted through the mould metal to the outer layer of water flowing through a cooling channel and then to the next inner layer. The sole mode of heat

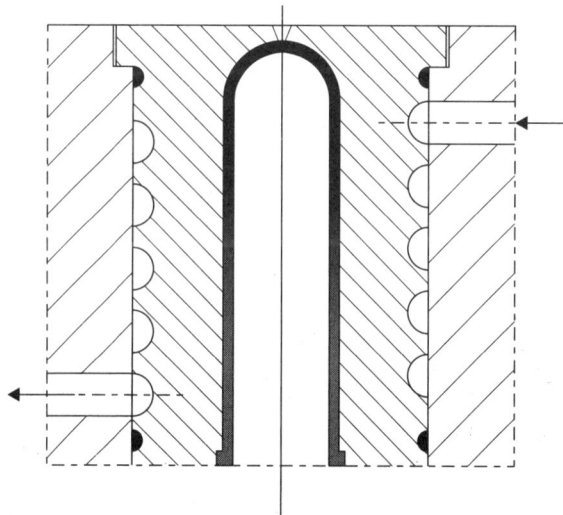

Fig. 4.39

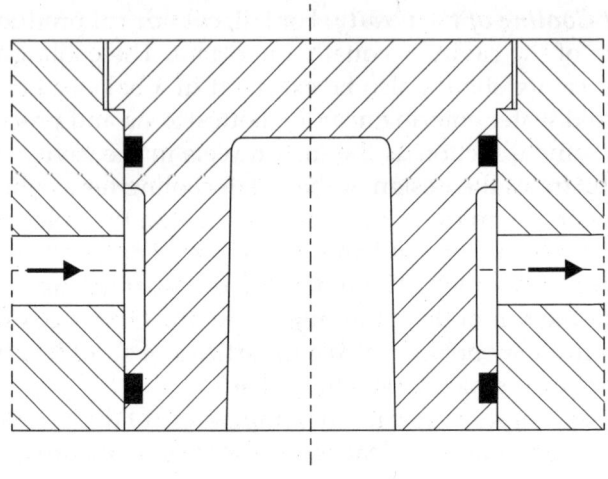

Fig. 4.40

dissipation is conduction, which is not very efficient due to poor thermal conductivity of water. However, if the flow is made turbulent (e.g. by introduction of spirally twisted metal strips in the cooling hole), water will be mixed constantly and heat dissipation will take place through conduction as well as convection.

Mould Wall Temperature

The ideal mould temperature is different for different thermoplastics. Formation of crystallites in case of semi crystalline polymers takes place well above their glass transition temperatures. The mould wall temperature for these polymers has to be maintained accordingly. It also facilitates good flow of the melt. The surface finish of the moulding is influenced by the mould wall temperature. Frozen-in stresses and warpage can also be minimised by maintaining the right temperature recommended for the polymer. The post-shrinkage will also be lower as well as more uniform with correctly cooled/heated mould. Indirectly, it contributes to the dimensional accuracy.

It is recommendable to cool or heat the mould through a temperature controlled unit as it makes it possible to re-create the exact moulding conditions when the mould is put in production again. The temperature of water, straight from a cooling tower, fluctuates with the ambient conditions.

Table 4.1 gives the most suitable mould wall temperature for various thermoplastics.

Table 4.1: Mould wall temperatures	
Polymer	*Temperature °C*
CA	20–50
LDPE/ HDPE	20–70
PP	20–80
PS	20–60
HIP	50–70
SAN	40–70
ABS	40–70
PVC	40–60
PMMA	50–80
POM	70–120
PC	80–110
PA 6	50–80
PA 66	60–90
PETP	70–110
PBTP	70–140
PPO	70–150
PPS	80–180
PES	120–190
PSU	140–190
PEI	135–165
PEEK	160–200
Polyblend PC+ABS	50–100
LCP	40–150

Ejection

When the injected melt cools in the mould, it shrinks in volume and hence becomes smaller in all dimensions. Consequently it becomes free from the cavity but holds onto the core, pins forming holes, ribs, bosses and other such details. After the mould opens, the product, sticking onto the core, has to be ejected by force. This force is applied onto the moulding by means of specific mould elements forming the ejection system. The guidelines governing the design of the ejection system and the choice of ejection elements are:

- The moulding should be ejected without deformation and damage
- The ejector marks should not be visible disturbingly.
- The ejection elements should not interfere with other functional components and features such as cooling or feeding

Ejection Elements

There are several standard ejection elements which are incorporated in the mould and serve to push out the solidified moulding. Their shape and design varies with the configuration of the moulded article. Although it is possible to compute the force holding the moulding, it suffices to choose surface area of the ejecting elements as one tenth of the area of holding, viz. that of the vertical walls, core pins, bosses, etc. The usual ejection devices are:

Ejector pin (Fig. 4.41): It is the most commonly used ejection element. Round ejector pins are available in standard diameters, ranging from 0.5 mm to 20 mm or more in various lengths in hardened state. Their incorporation in the mould is very simple; a drilled and reamed H7 hole in the core or sometimes the cavity, enlarged after a certain guiding length, is practically all that is needed. Their lower end culminates in a flat head of bigger diameter, which serves to hold them in the ejector plates as shown in the introductory mould design Fig. 4.1. They are basically suitable for ejection of shallow mouldings made of rigid/semi-rigid materials. They leave a mark on the surface of the article. Round pins are also employed to eject the cold slug and the runners.

Blade ejector (Fig. 4.42): This is a variation of the round ejector pin and resembles one with certain length in front machined down flat. It is employed to eject mouldings with thin walls and deep ribs. The aperture for a blade ejector in the cavity is

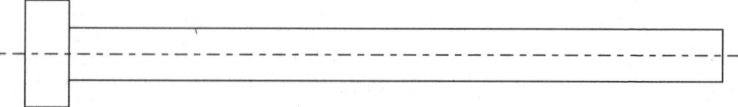

Fig. 4.41

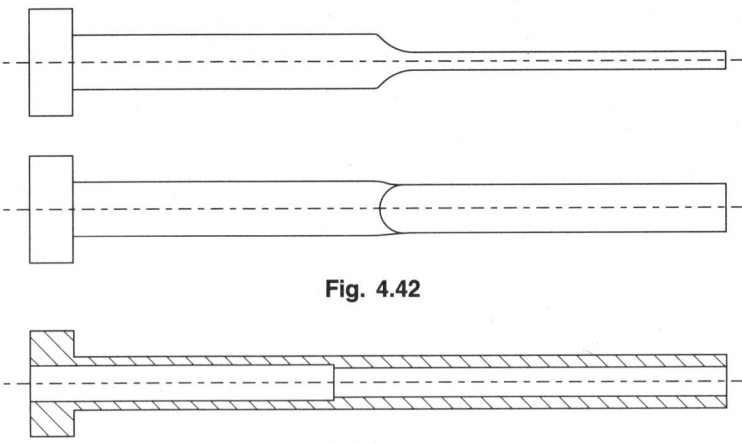

Fig. 4.42

Fig. 4.43

made by EDM, wire cutting or by parting the cavity insert and milling a rectangular slot for the ejector at the joint.

Sleeve ejector (Fig. 4.43): It resembles a round ejector pin with a round bore in the centre along the longitudinal axis. Sleeve ejectors are employed for ejection of round articles and for those having bosses with holes formed by round core pins. They exert a uniform, alround ejection force for distortion free demoulding and do not leave a visible mark when used under a round boss.. Because of the accuracy required both on outside as well as inside diameter, a sleeve ejector is expensive. Sleeve ejectors are available ready made in some standard sizes.

Figure 4.44 shows various modes of the application of above ejectors.

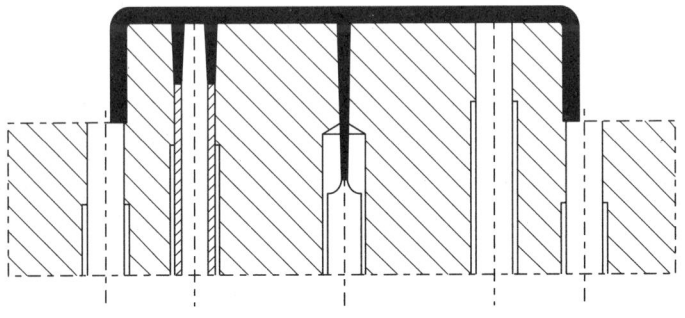

Fig. 4.44

Mushroom ejector: Tall products, especially those out of softer plastics, cannot be demoulded with the help of ejector pins without distortion and damage. The disposable thin walled tumblers are one such example. Here a specially designed mushroom ejector ensures safe ejection as it distributes the applied force over a large area. It constitutes a part of the core and must be foreseen with cooling. In appearance, it resembles a poppet valve (Fig. 4.45).

Mushroom ejector is usually anchored in the ejector plates and moved by them.

Stripper ring/stripper plate: For articles having a regular, geometrical contour and relatively deep walls such as tumblers, round or rectangular containers, tubular products, bowls and even technical components, where the main holding is along the walls, stripper ejection is the most suitable method for safe demoulding. A closely fitting ring around the core, which sits under the moulding and is connected to the ejector plates, strips the cooled moulding off the core after the mould has opened (Fig. 4.46). It is equally suitable for soft as well as rigid plastics and does not leave a mark on the surface. Its action is the same

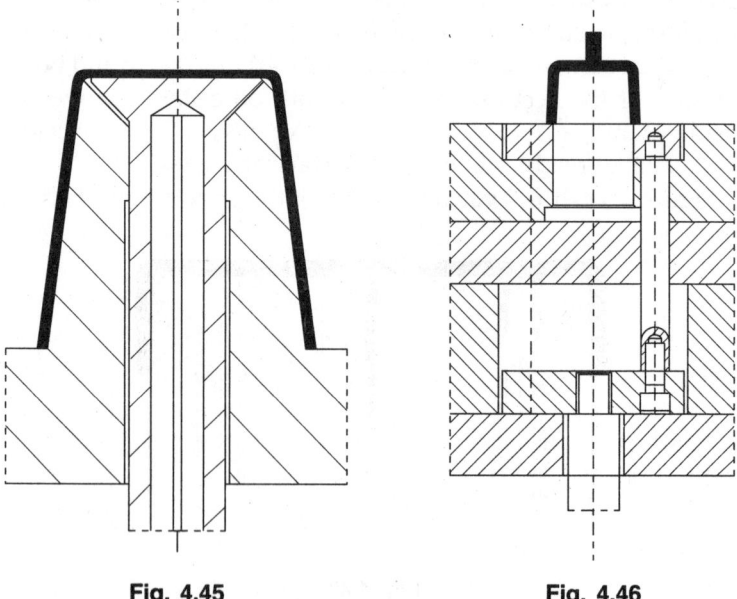

Fig. 4.45   **Fig. 4.46**

as that of a sleeve ejector. In case of moulds for large articles or those with a number of cavities, a stripper plate instead of stripper rings may perform the demoulding action. Hardened rings may be fitted in an unhardened mould plate, which is activated by the ejector plates.

The stripper plate, as mentioned before, is generally connected with the ejector plates of the mould and pushed by them during the ejection operation. Alternatively, the stripper plate can be pulled by bolts attached to the fixed mould half, which come into action after the mould has opened by a predetermined distance (Fig. 4.47). The method is economical as it obviates the need of ejector plates. It is also independent of the stroke of the machine ejector.

The method is also applicable to a stripper plate situated on the injection side.

Stripper bar (Fig. 4.48): It is a variation of the stripper ring and is employed to demould articles with a partially geometrical

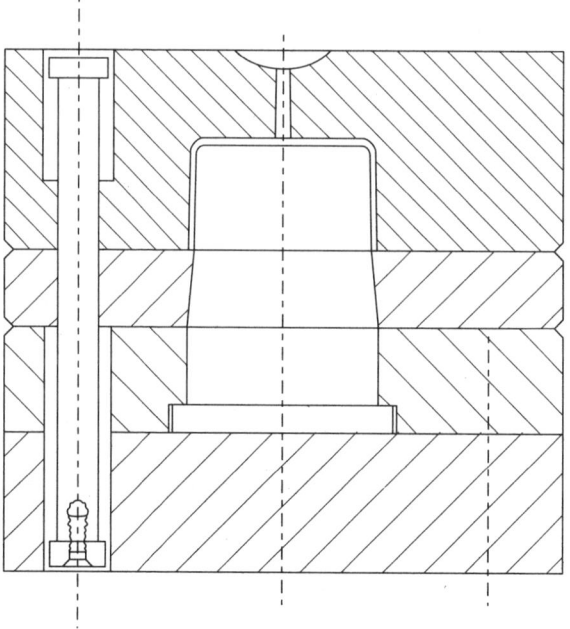

Fig. 4.47

contour. Stripper bars are located under the straight edges of the moulding and moved through connecting bolts joining them with ejector plates.

Stripper bars may be employed in conjunction with other ejection devices such as ejector pins, ejector sleeves etc.

Air ejector: Articles with geometrical contours such as buckets, bins and big containers, usually moulded out of semi-rigid plastics such as polyethylene, and polypropylene, could be demoulded by stripping. Their size, however, would call for huge stripper rings and extremely long ejection strokes. Air ejectors (Fig. 4.49) are round air valves placed under the bottom of the product. Compressed air pushes them up by distorting the moulded bottom in the elastic range and fills up the gap under the entire bottom.

The force of compressed air working upon whole area is sufficient to release the moulding from the core. It may also happen that the moulding sticks in the cavity due to the vacuum. The force of suction may be greater than that of

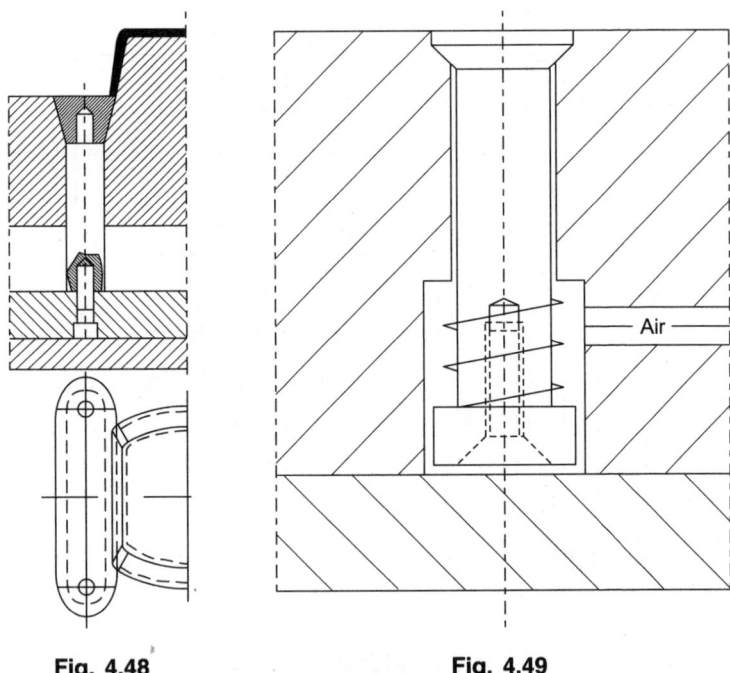

Air

Fig. 4.48   **Fig. 4.49**

shrinkage. To break the vacuum, air ejectors may also be placed on the cavity side. Moulds for buckets usually have this provision. However, the working principle remains the same as discussed above.

Another configuration of air ejector, for use in cases were the ejector mark may not be desirable, is shown in Fig. 4.50.

Choice and Location of Ejection Elements

Mouldings, sticking onto the core by dint of shrinkage of the plastics, are ejected with the help of demoulding elements described above. The placement as well as the size of ejectors plays a vital role in safe, distortion free demoulding.

The vertical walls of the core are the main site of the holding force. Ejectors should be placed as close to these sections as possible. Ejector pins can be used with rigid plastics but with flexible or semi-rigid materials like LDPE, plasticised PVC, EVA, etc. the ejector pins, if unavoidable, should be situated almost in line with the vertical walls. As a rule, other elements of ejection such as stripper rings and bars, sleeve ejectors and

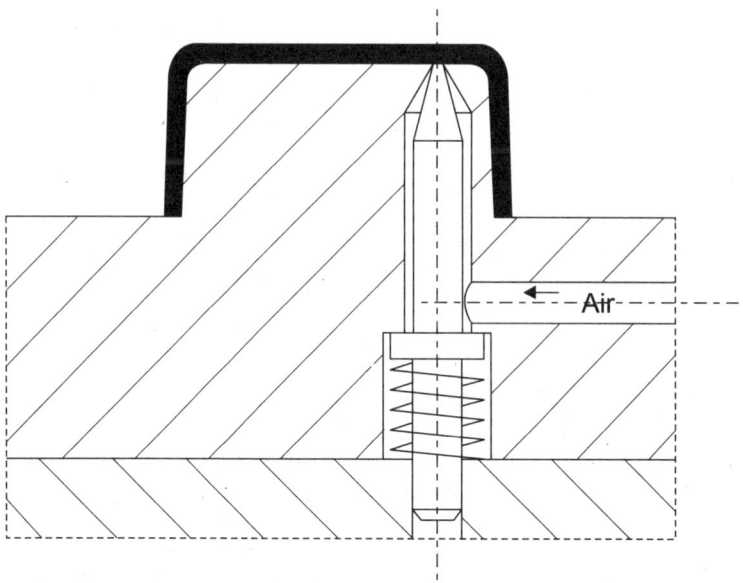

Fig. 4.50

mushroom ejectors should be given preference for the semi-rigid plastics mouldings.

Ribs bosses and hole forming pins also hold the moulding and should be foreseen with suitable ejectors.

The vertical walls of the core should be free of machining marks and should be polished in direction of ejection. It may be remarked that some plastics tend to adhere to polished surfaces. In such cases, the surface is roughened through fine sand blasting after polishing as recommended. Taper facilitates ejection.

Undercuts

The means of ejection considered so far apply to mouldings, which hold onto the core or cavity by virtue of shrinkage or vacuum. Some articles, however, may have design features, which cause holding through interference in the path of ejection. These details, preventing direct ejection, are known as undercuts and may be on the inner or the outer side of the article. Side holes across the axis of ejection, continuous or interrupted grooves and beads around the circumference, internal and external threads, raised or depressed lettering, logos and designs on the side walls are some of these features. Obviously, if the moulding is pushed forward in the conventional manner before removing the obstacles, it is bound to get damaged. In some cases, the mould may suffer too. One must, therefore, make arrangement to remove these hindrances before the ejection process can be actuated.

Figure 4.51 shows an example where aperture at an angle to the line of ejection has been neutralised by splitting the obstruction and dividing it between the core and the cavity. Here too, a section of the parting line has been shifted so that all points of the moulding are in the line of draw.

Mouldings with external undercuts like threads, grooves and projections (Fig. 4.52) can be forced out of the cavity safely, provided these features are well rounded and dimensionally within a certain limit, depending upon the rigidity of the plastics in question (Table 4.2). It is, however, essential that the mouldings are in a position to collapse inwards elastically. In order to facilitate that, the internal core must be pulled out before the ejection process sets in. In other words, the ejection process has

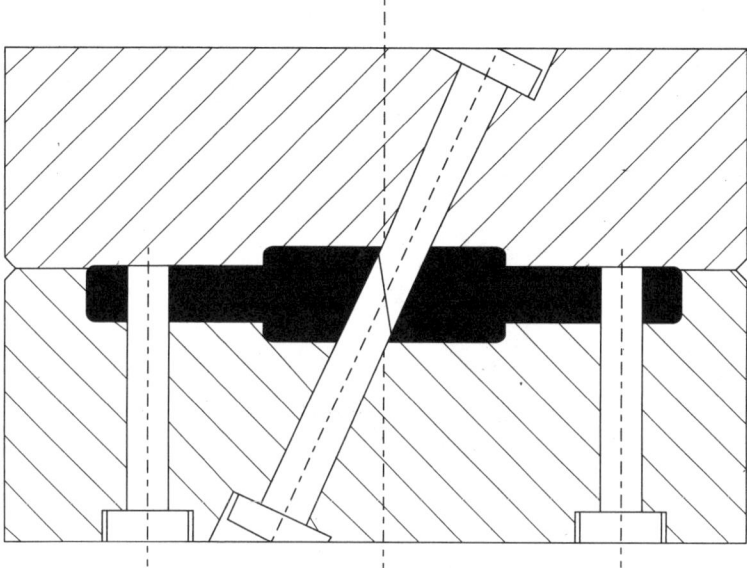

Fig. 4.51

to be performed in two steps, viz. pulling out the core and then pushing out the moulding. It is referred to as two stage ejection

Two Stage Ejection: As described above, demoulding may have to be carried out in two steps in certain cases depending upon the design of the article. It may be necessary to move one or more mould plates before the final ejection.

The classical device of moving a mould plate for a predetermined distance and then releasing it is illustrated by Figs 4.53a to c. It consists

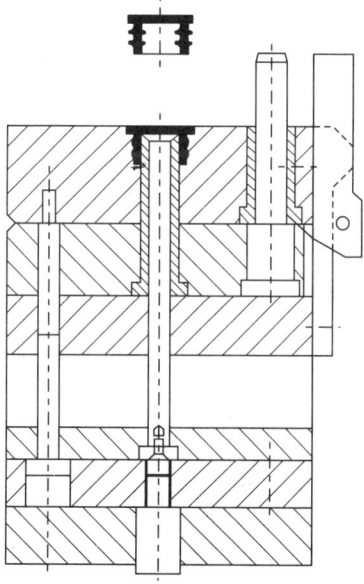

Fig. 4.52

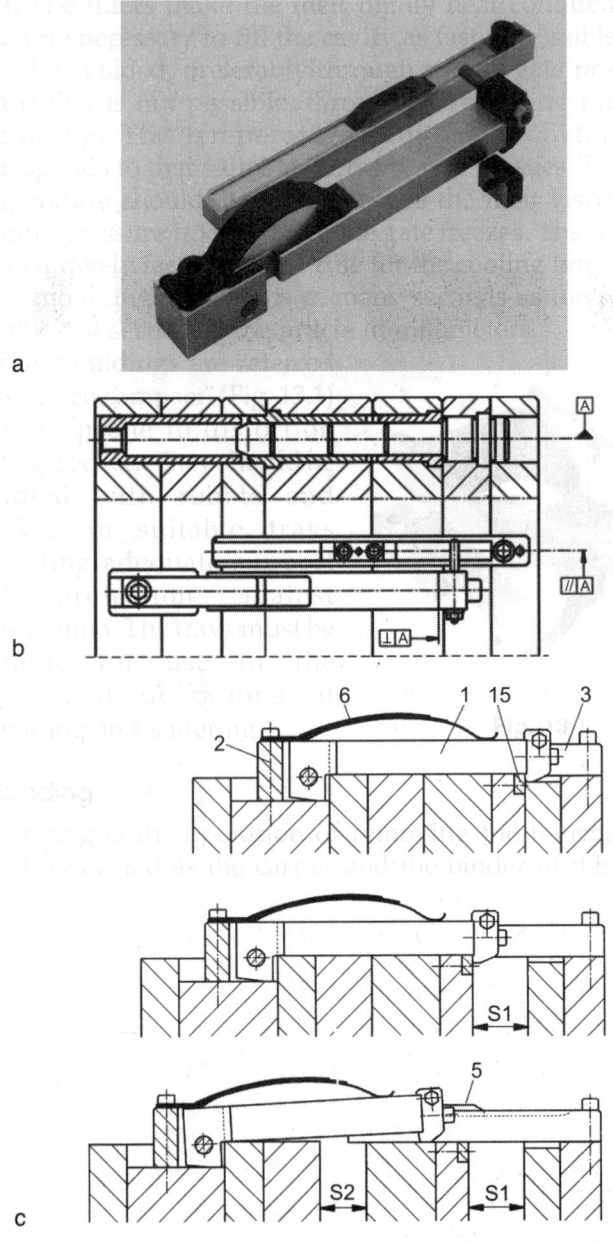

Figs 4.53a to c

of a hinged claw and a cam plate with an adjustable hump. The claw is fixed onto the moving mould half and holds the loose plate. The cam plate is fixed onto the stationary mould half. As the machine opens, the moving mould half moves away and takes along the loose plate held by the claw till the hump of the cam forces open the claw and sets the plate P free. It does not move further with the ejection mould half.

It may be noted that in some cases, it may be found necessary to reverse the position of the two components of the device. The claw may be fixed on the stationary side and the cam plate onto the moving half. The working principle, however, remains unchanged. The device is employed in pairs to apply force on both sides of the plate.

There are numerous solutions to the central idea of two stepped movement. Several devices to execute the movement, including the one shown above, are available as standard mould components.

Such devices are also employed in three plate moulds to move the cavity plate and create gap for the runners to fall.

External Undercuts

Side holes, external threads, grooves, beads, projections and depressions on the outside of an article are usually cleared by making the mould component forming the undercut move outwards across the axis of ejection before the ejection device is set in operation. The cross movement of the sliding mould component can be achieved either through mechanical devices or with the help of pneumatic or hydraulic cylinders.

The slider with slanting pin. The most common mechanical mode of converting the available opening stroke of the machine into a movement across is shown in Fig. 4.54. A cylindrical pin, fixed in one half of the mould at an angle, passes through the moving mould component, which is retained by guides fixed on the other half of the mould. As the mould halves move apart, the slanting pin imparts a cross movement to the slide. The extent of the stroke is decided by the angle and the length of the slanting pin. The angle of inclination with the mould axis should not exceed 30 degrees as otherwise the component of the force acting across it ma©y bend and damage the pin.

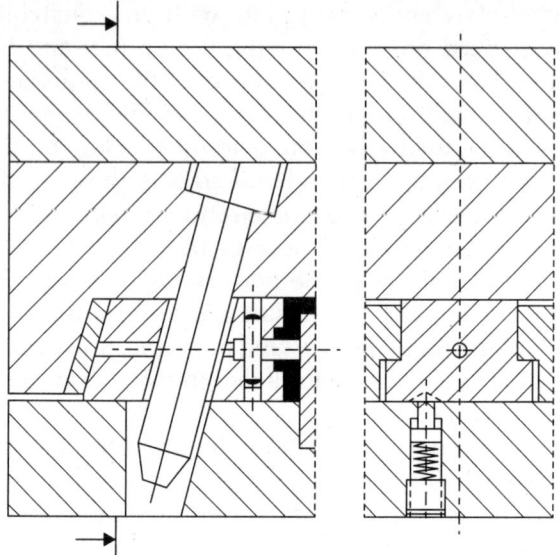

Fig. 4.54

The force exerted by the melt on the slide is countered by the angular heel formation in the plate housing the pin. It backs up the slider during injection. It may also be shaped as a separate hardened insert, fixed in the plate. The angle at the end of the moving block and consequently on the face of the heel should be at least as much as the inclination of the pin to eliminate fouling. In practice, it is kept about 2–3 degrees larger to be on the safer side.

The cross movement is repeated in reverse direction when the mould closes. Again the movement is imparted by the pin only; the heel comes in contact with the slide at the fag end. Hence there is no friction and wear which could lead to clearance and to flash.

It is absolutely necessary that the slide stays at the position where it was left by the pin so that the later can re-enter the bore when the mould closes. A spring-loaded ball catcher, as shown in the diagram, arrests the slide at the end of its travel. The device works satisfactorily if the slide moves horizontally. A slide moving up or down, viz. vertical to the machine axis may drop down or change its position due to its own weight which may exceed the force exerted by the catcher spring. Unless unavoidable, the mould layout should foresee a horizontal movement of the slides. The vertically moving slides should be retained in the end position with more positive arrangements like end stops and springs.

Depending upon the shape of the product, the outer contour may be partially or wholly formed by slides. The later case is referred to as a split cavity.

The slider with dog leg is an improved version of the slanting-pin mechanical device to achieve movement at an angle to that of the machine. It has the advantage that the starting point of the slide movement can be delayed as much as required (Fig. 4.55). Obviously, it is more expensive in execution.

The slider with cam plate (Fig. 4.56): The cross movement in this version of the sliders is generated by a cam plate fixed on injection mould half which guides a pin fixed with the slider block. The slider pair covers the whole length of the core plate. It is generally employed when the outer shape of the product

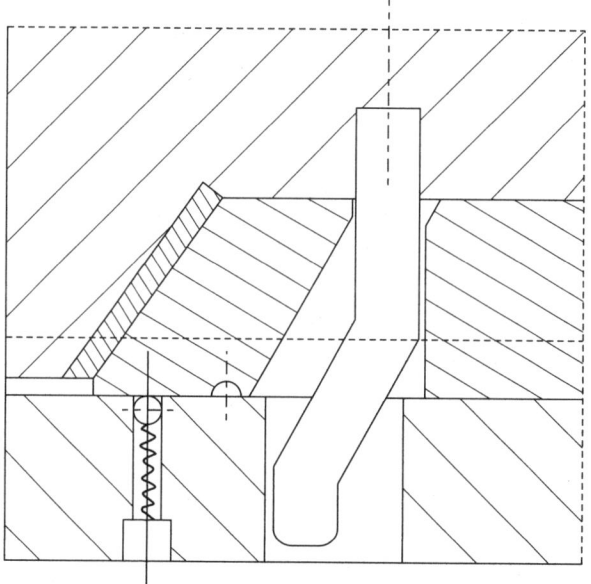

Fig. 4.55

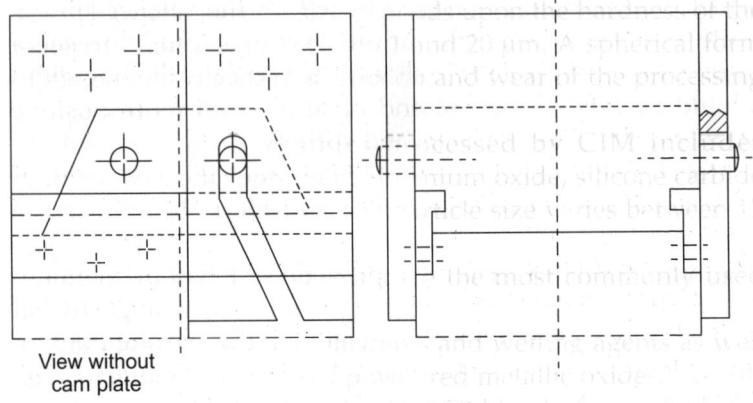

View without
cam plate

Fig. 4.56

has to be formed by means of splits. The cavities are placed in a row.

The design enables use of the whole length of the slider. One or more cooling lines can be easily incorporated in the block.

Pneumatic/hydraulic cylinders: Mechanical devices for cross movement are subject to limitations posed by the machine stroke. Pneumatic or hydraulic cylinders can deliver the required travel independent of the machine movement. They can be operated at any juncture during the opening stroke of the machine. The sliders fulfil the same function as described before and do not differ in design, except that they are directly connected to the cylinder providing the motive force. The cylinders are fixed to the mould half housing the slider. Self locking cylinders do not need additional mechanical backing for countering the injection force.

Internal Undercuts

Internal threads, grooves, beads, and blind pockets are some of the undercuts, which come in the way of ejection. Such obstructions, if not very deep and sharp, can be overcome by sheer ejection force, provided the outside of the product is free before ejection to expand within the elastic limits and skip over the core forming the obstruction, analogous to clearing of external undercuts illustrated in Fig. 4.52. The main criteria is the yield stress at the ejection temperature. The undercuts must be well rounded to facilitate slipping over the projection. Flexibility of the plastics used is also an important factor. The bigger the article and the more flexible the plastic, the deeper can be the undercut.

The position of the undercut is crucial as an undercut near the open end of a container will release more easily than the one close to the bottom. For the later case, half the permissible depth will be safer.

Table 4.2 lists the amount of expansion for safe ejection for various plastics.

Table 4.2: Permissible elastic expansion	
Material	*Max. Elastic Expansion (%)*
PS	0.5
SAN	1.0
ABS	1.5
PC	1.0
PA	2.0
POM	2.0
LDPE	5.0
HDPE	3.0
PP	2.0
PVC rigid	1.0
PVC plasticised	10.0

Figure 4.57 illustrates an unusual application of this attribute to form grooves for flexible seals in moulded sanitary fittings. The internal mould core A, forming the groove, is extracted after the outer cavity inserts C, restricting expansion of the pipe end, have been pulled away. The end of the fitting can expand within elastic limits and regain its moulded shape.

Tapered Slides: Partial internal undercuts, interrupted protrusions like partial internal threads, etc. are cleared by making the relevant mould insert or the tongue forming the undercut move forward at an angle during the ejection stroke (Fig. 4.58) Depending upon the angle, the undercut is made free after a particular distance. The product can now be ejected in the usual way.

Large screw-on closures are often provided with interrupted internal threads so that they form two segments. The method to release the undercut is to make the mould component

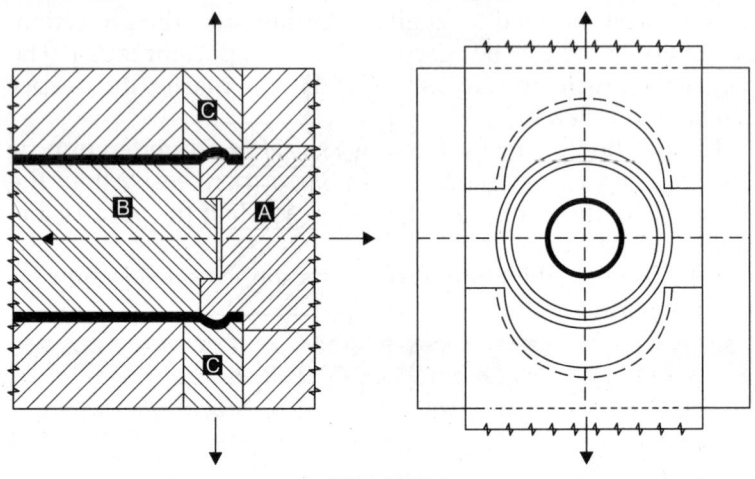

Fig. 4.57

containing the threads as tapered slides (A) and the unthreaded middle section (B) as a central core with corresponding slope.

The three components together form the mould core when the mould is in closed position. The cores with threads are housed in the stripper plate whereas the central core is fixed in the plate below. Figure 4.59 depicts the arrangement.

As the mould opens, the stripper plate is moved forward, leaving the central core (B) behind and creating a gap for the threaded cores (A) to move inwards. After some distance the moulded threads are free of the cores. The mouldings can be ejected.

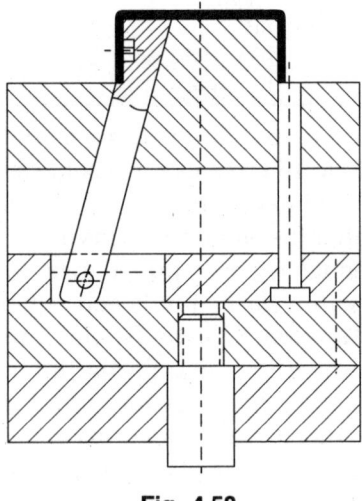

Fig. 4.58

Collapsible Core: Another device, well suited for clearing continuous internal undercuts like threads, grooves, projections, etc. is the collapsible core. Figure 4.60 depicts its working

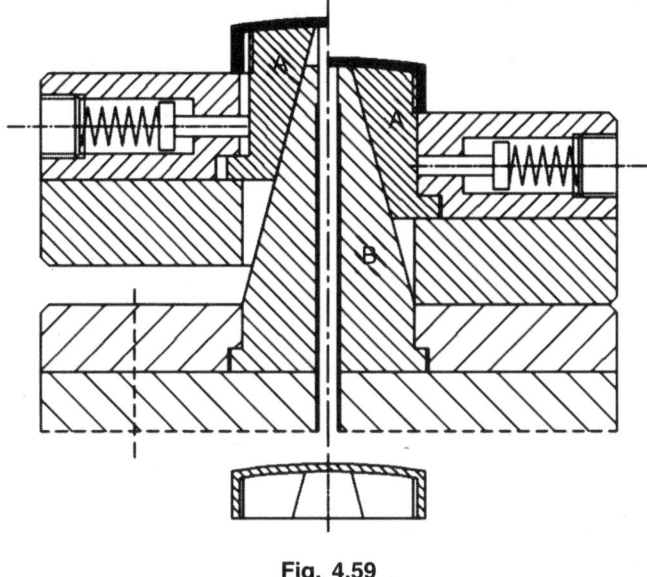

Fig. 4.59

principle. The device consists of two main components; the stationary internal core and two sets of sliding segments, arranged around the stationary core. The segments, made of spring steel, form a complete circle in cross-section when the tapered inner core is in the forward position. They collapse inwards to their original position when the core is withdrawn.

The sketches 1 and 2 illustrate the working principle of the collapsible core. The central core is stationary whereas the segments slide forward with the plate housing them. The segments move inwards, i.e. "collapse" as space is created

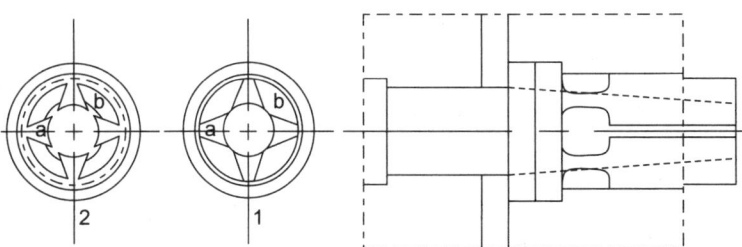

Fig. 4.60

through withdrawal of the tapered core. Thus the undercut gets cleared.

Another version of the collapsible core attaches the segments to the tapered central core by means of dovetails which are cut at different inclinations. The inward movement of the segments is now dictated by the dovetails and not by the natural resilience of the spring steel segments.

As the angles of inclination for both sets of segments are different, they do not clash and hinder each others movement.

Unscrewing of Internal Threads: The most common internal undercut in mouldings is the internal thread. Screw caps and closures with internal threads for bottles and containers are produced in large quantities. Closures with round threads can be stripped off but other forms of threads call for some safer device to clear the undercut. In other words, the moulding must be unscrewed. The methods of unscrewing commonly resorted to involve:

- Holding the moulded article to prevent rotation
- Rotation of the threaded mould core
- Receding of the core during rotation or
- Lifting up the threaded moulding during rotation of the (non-receding) core.
- Ejection.

The most common method is to rotate the threaded cores in the mould, The rotation can be effected manually, mechanically, hydraulically or electrically. The mouldings like screw caps are provided with notches on the periphery around the threads to prevent their simultaneous rotation along with the cores (Fig. 4.61).

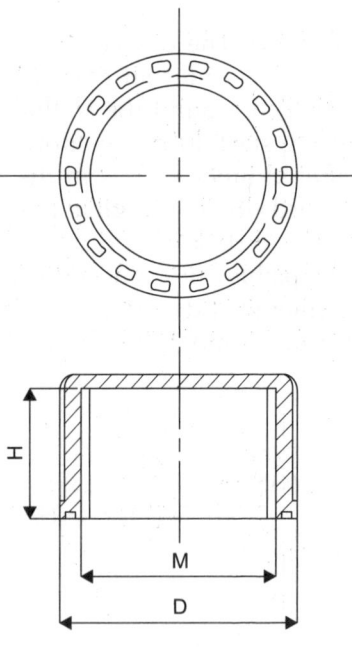

Fig. 4.61

Numerous methods of unscrewing have been developed. A few typical designs will be reviewed here.

Manual Method: Adopted for very small quantities and mostly with single cavity hand moulds, the method consists of inserting a loose core with the threads in the mould and ejecting it with the moulded closure. For the next shot, another identical core is laid in the mould. The ejected moulding is unscrewed from the first core manually.

Semi-automatic unscrewing: It is a mechanical method used with small moulds, especially when the quantity required is not very large. As obvious from Fig. 4.62, the rotary movement executed by the operator in horizontal plane is transferred to the threaded core of the mould through the bevel gears. The mould core does not recede but the ribs on cap's circumference prevent it from rotation and make it lift up and come out of the cavity.

With appropriate modification, the method can be converted into an automatic one.

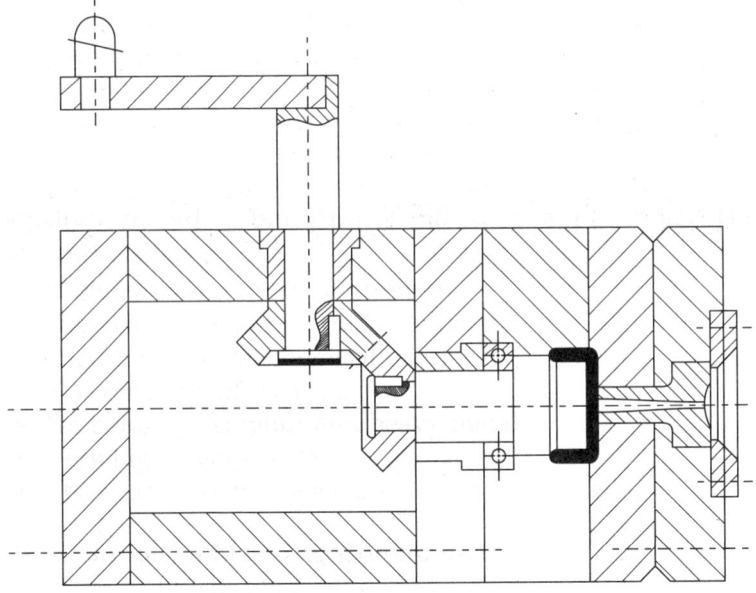

Fig. 4.62

Fully automatic unscrewing: The operation of unscrewing can be made automatic if the threaded mould cores are rotated automatically. The rotational movement can be generated by:
- Converting the linear movement of platen into rotation during the opening stroke of the moulding machine.
- Hydraulic cylinders
- Hydraulic or electrical motors

Section 4.2 on Special Mould Designs describes and depicts examples of automatic unscrewing employing the above mentioned methods.

The Ejection Force

By dint of shrinkage, the mouldings tend to become smaller but are hindered by the mould core. Consequently they hold onto the core with a force which is proportional to the amount of shrinkage and the elastic modulus of the plastics, as discussed before. The ejection force has to overcome the frictional force between the moulding and the core. Understandably, the frictional force exists on the vertical sections, where the moulding has to move in the direction of ejection. It is a product of the holding force caused by the shrinkage and the coefficient of friction between the particular plastics and the metal of the mould core (Table 4.3). The ejection force is applied on the moulding through the ejection elements described before.

Table 4.3: Coefficient of friction; plastics on steel	
Thermoplastics	*Coefficient of friction*
LDPE	0.40
HDPE	0.25
PP	0.33
PS	0.40
SB	0.50
SAN	0.50
ABS	0.50
PVC	0.55
PMMA	0.40
POM	0.25
PA 6	0.40
PC	0.55

The coefficient of friction can be reduced by improving the surface of the mould core through grinding, polishing, plating, coating etc.

It becomes a complicated task to calculate the force of ejection in case of irregular, ungeometrical shapes. A thumb rule, which has been found to be quite close to the mark, is to provide ejectors with total surface area equalling one tenth of the area of vertical walls of the core.

The Unscrewing Force

As discussed before, mouldings hold onto the core with certain force. It is proportional to the area, the amount of shrinkage, modulus of elasticity of the particular plastics and the coefficient of friction between the plastics moulding and the metal core. In order to unscrew a moulded core, a torque has to be applied to overcome the holding force. It can be computed as explained below.

Assuming for the illustrated screw cap (Fig. 4.61):

Shrinkage $\quad\quad\quad\quad s(\text{mm}/\text{mm})$
Elastic modulus of plastics $\quad E(\text{N}/\text{mm}^2)$
Coefficient of friction $\quad\quad f$
D = The outside diameter of the cap
M = The diameter of threads
H = The height of threads
A = Area of threads = $\pi \times M \times H$
t = Wall thickness = $(D - M)/2$
The circumferencial stress St = Exs
The pressure exerted by shrinking moulding on the core
$P = \text{St} \times 2t/M$

Shrinkage force $\quad\quad$ Fs = $P \times A$
Total frictional force $\quad$ Fr = Fs $\times f$
Torque $\quad\quad\quad\quad$ Tr = Fr $\times M/2$

It may be remarked that the friction is reduced significantly if the threads are well machined and subsequently ground, polished and/coated.

Mould Shrinkage

Plastics expand in volume when melted and contract when cooled. During injection, they are in the molten and expanded state and occupy more space. Hence a shrinkage allowance has to be incorporated while deciding the mould dimensions.

Mould shrinkage is defined as the ratio of dimensional difference between the moulding and the mould, measured 24 hours after moulding, at room temperature. Dimensions of the mould for a proposed article have to be chosen larger by the amount of expected shrinkage. In mathematical form, the mould shrinkage may be expressed as:

Mould dimension – Article dimension

Mould shrinkage (in %)

$$= \frac{\text{Mould dimension} - \text{Article dimension}}{\text{Article dimension}} \times 100$$

Mould shrinkage is not an absolute factor which could be applied to all dimensions uniformly. It is always stated as a range for individual plastics. The amount of shrinkage depends upon the following factors:

- Amorphous polymers expand less in volume (~15%) than the semi-crystalline ones (>30%) upon melting. Consequently, the former shrink less than the later while solidifying.
- Semi-crystalline thermoplastics have greater shrinkage in direction of flow than perpendicular to it. The amorphous ones exhibit a more uniform shrinkage.
- Higher injection and holding pressures decrease the mould shrinkage.
- Higher melt temperature leads to higher shrinkage.
- Higher mould temperature increases the shrinkage.
- Restricted dimensions, such as the diameter around a core or the distance between two hole-forming core pins, do not shrink till the moulding is ejected. Their final shrinkage outside the mould is less than that of the free dimensions.
- Thicker sections shrink more than the thinner ones.
- Filled thermoplastics shrink less than the unfilled ones. The difference in shrinkage also depends upon the degree of orientation of the fibres.
- Some plastics undergo post-shrinkage over a long period. Again, some hygroscopic polymers like polyamides, on the other hand, expand after ejection after they have absorbed moisture.

- The expansion of mould dimensions, when the mould is heated, must also be accounted for when computing the shrinkage allowance.

The mould designer must take into account these conditions while dimensioning the core and the cavity of a mould in design. The gate position, especially with the semi-crystalline plastics, also influences the extent of shrinkage as it decides the direction of flow and where the material freezes last. Sections which remain longer under holding pressure shrink less.

Table 4.4 lists the mould shrinkage of various thermoplastics processed by injection moulding.

Table 4.4: Mould shrinkage		
Acronym	*Polymer*	*Shrinkage %*
CA	Cellulose Acetate	0.4–0.7
PS	Polystyrene	0.4–0.6
SB (HIP)	High Impact Polystyrene	0.4–0.7
SAN	Acrylonitrile Styrene	0.4–0.7
ABS	Acrylonitrile Butadien styrene	0.4–0.7
PVC-R	Polyvinyl Chloride Rigid	0.4–0.5
PVC-S	Polyvinyl Chloride Soft	1.5–2.5
PMMA	Polymethyl Methacrylate	0.4–0.8
PC	Polycarbonate	0.6–0.8
PPO	Polyphenylene Oxide	0.4–0.7
PSU	Polysulfone	0.7–0.8
PETP (amorphous.)	Polyethyleterephthalate	0.2–0.4
PETP (linear)	Polyethyleterephthalate	1.2–2.0
PBTP	Polybutyleneteraphthalate	0.7–2.2
LDPE	Low Density Polyethylene	1.5–3.0
HDPE	High Density Polyethylene	1.5–4.0
PP	Polypropylene	1.0–2.5
PA 6	Polyamide 6	0.8–2.1
PA 6 6	Polyamide 6 6	2.5–3.1
PA 6 10	Polyamide 6 10	1.2–1.6
PA 11	Polyamide 11	0.8–1.2
PA 12	Polyamide 12	0.7–1.1
POM	Polyoxymethylene	1.9–2.3
PEEK	Polyether Etherketone	0.7–1.2
LCP	Liquid Crystal Polymer	0.3–0.6

The shrinkage figures, as listed above, are average values. These may be different for the same polymer from different manufacturers. It is advisable to go by the data provided by the particular supplier.

Tolerances

Tolerance is the total permissible variation of size, form or location. The aim is to ensure that the fit, function and performance of the moulded component are not effected in spite of the variation.

Tolerances or the extent of dimensional fluctuations of thermoplastics mouldings are dependent upon the following factors:

- The polymer and its shrinkage range
- The moulding parameters
- The melt and mould temperature
- The pressure and its duration
- The constancy of machine settings
- The constancy of polymer characteristics, size of granulate, viscosity
- Constancy of the ambient conditions

Shrinkage of certain polymers varies within a broad band. Tolerances or the variation of dimensions for products moulded out of these plastics would naturally be more than those for polymers with less shrinkage.

Moulding parameters, especially the pressure, temperature and time, influence the shrinkage and consequently control the dimensional variation of the final product.

Material and mould temperatures directly influence the flowability, the filling time, the injection pressure, the effective period of holding and the cooling time. Any variation in these temperatures effects shrinkage and consequently the tolerance.

The degree of constancy and accuracy of machine settings, viz. the temperature, pressure, time of different operations and size of the melt cushion, influences the dimensions of the mouldings directly. Dimensions of mouldings from a closed loop machine will be in a narrower range of variation than of those moulded on a conventional machine.

The melt flow index of a polymer may vary from batch to batch. A conventional moulding machine cannot adjust itself accordingly. The consequent change in pressure and speed of injection also leads to a change in dimensions of the product.

The ambient conditions such as the temperature, humidity and air circulation at the place of moulding influence temperature of the barrel, the mould, and the hydraulic oil which effects pressure, injection speed, rate of cooling, etc. Unless kept constant, they can lead to wide variations in final dimensions.

It will be seen in the light of above facts that a uniform standard of tolerance cannot be applied to all materials and all cases. It is advisable to limit the narrow tolerances to as few dimensions as possible. Tool makers tolerance should also be taken into account.

Table 4.5 gives the practicable tolerances for open dimensions with unspecified tolerances of moulded articles.

Table 4.5: Dimensional tolerances	
Dimension in mm	*Tolerance in mm*
1–6	±0,1
6–10	±0,2
10–18	±0,3
18–30	±0,4
30–50	±0,6
50–80	±0,8
80–120	±1,0
120–180	±1,3
180–250	±1,6
250–315	±2,0
315–500	±2,6

The Parting Line

The mating face of the cavity and core is known as the mould parting line. The mould opens at the parting line to enable extraction of the cooled moulding. The shape of the product is one of the major considerations governing the course of the parting line. The main criteria for the choice is the safe ejection of the moulding.

The parting line for most articles with geometrical configuration is usually a straight line in one plane. In other words, the mating surface of two mould halves is flat. In many such moulds, the cavity for the product is situated on one side only (Fig. 4.63). The plane parting line, however, may have to divide the cavity in two parts to enable extraction (Fig. 4.64). Such a parting line calls for utmost accuracy in matching of the two halves.

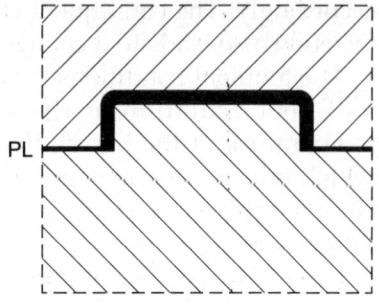

Fig. 4.63

Depending upon the outer contour of the product, the parting line may follow a straight, a slanting, a stepped or a curved path. It may even be a combination of these (Figs 4.65 to 4.67). As a rule, the parting line passes through the highest points on the periphery of the moul-

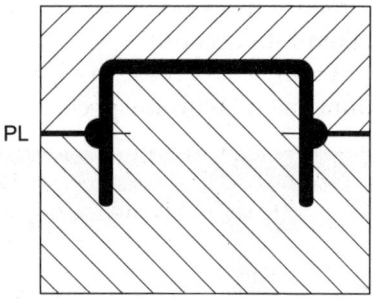

Fig. 4.64

ding to circumvent undercuts. Industrial components, products with free curves, toys, etc. necessitate choice of such parting lines. Parting surface for animal or human figures may even be three dimensional.

Apart from the product configuration, the other factors having a say in the choice of the parting line are:

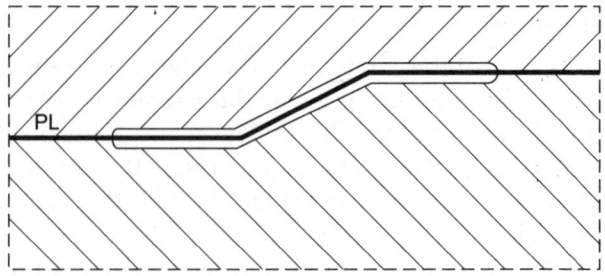

Fig. 4.65

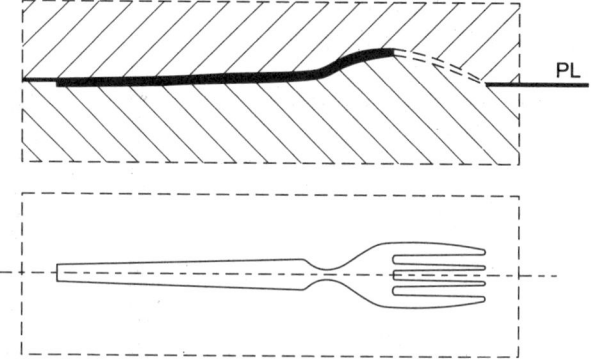

Fig. 4.66

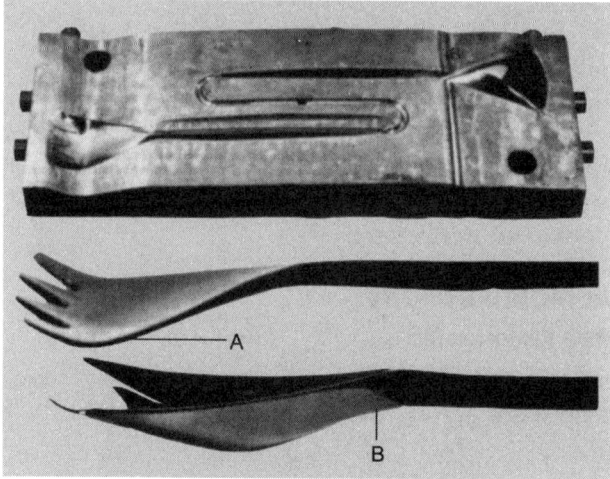

Fig. 4.67

- The polymer. The flexibility of some plastics permits ejection of undercuts like grooves, beads, threads etc. without having to part the cavity. The course of the parting line may be different with rigid or brittle plastics.
- The number of cavities. The mode of placement of the product and the course of the parting line in a multi-cavity mould may have to be chosen other than that in a single cavity mould in order to use common splits, slides, cores or other mould features.

- Aesthetic considerations. A comparatively less convenient parting may have to be chosen instead of a more favourable one in order to avoid marks on an aesthetically important surface of the moulding.
- Machine constraints. The product may have to be placed and parted differently than the obvious course in order to suit the machine dimensions. For example, a long cone placed conventionally in the mould calls for an opening stroke of at least

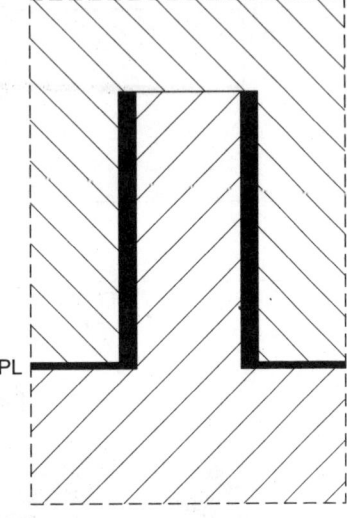

Fig. 4.68

twice its height (Fig. 4.68). Parted differently, it may be moulded on a machine with smaller stroke (Fig. 4.69).
- Venting. The parting line should facilitate free escape of the air contained in the cavity.

SPECIAL MOULD DESIGNS

The Three Plate Mould

In the so-called three plate mould, the mould cavities and the runner network are at two different levels so that the mould

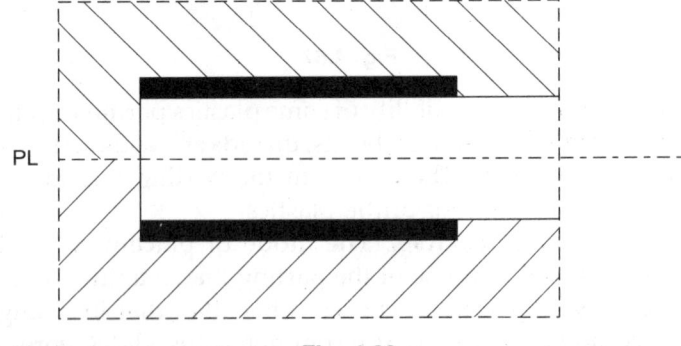

Fig. 4.69

has to open at two planes unlike a two plate mould. The mould has three distinct units; the fixed half, the floating cavity plate and the moving mould half. The runner network is situated between the fixed mould half and the cavity plate, at the back of the later. The sequence of operations is as follows:

1. As the machine opens after cooling of the moulding, the floating cavity plate remains attached with the moving mould half and moves away from the fixed half, thus creating enough space for the runners to fall. The runner is held onto the fixed mould half.

2. The mould opens further. The floating cavity plate moves on and the runner is stripped off from its holders. The cavity plate stops. The machine keeps on opening. The moulding sticks onto the core in the moving mould half.

3. The machine opens to its end position and the ejection system for the moulding is activated.

It implies that the mould requires additional devices to pull along the cavity plate by a predetermined distance. It also needs arrangement to hold the runner during the above operation and to strip it off at some juncture.

The arrangement adopted here makes use of the standard mould component called puller (see Fig. 4.53) discussed in the section "Ejection". The claws hold the cavity plate on the moving side during the opening stroke till cams disengage the plate by pushing back the claws.

As shown in the mould design (Fig. 4.70), the runner is held on to the fixed mould half by means of pins having undercuts. These runner-holding pins pass through a stripper plate. After the cavity plate has travelled a predetermined distance, it pulls the stripper plate by means of the distance bolts and the runner network slips over the holder pins. It can fall down by dint of its weight.

The additional plates and other paraphernalia for effecting the second opening to eject the runners make a three plate mould costlier and slower. The runner network, especially when the number of cavities is large, is quite unwieldy. It may tilt while falling and get entangled between the plates, endangering automatic operation of the process. However, it has some advantages too. It enables positioning of the gate at

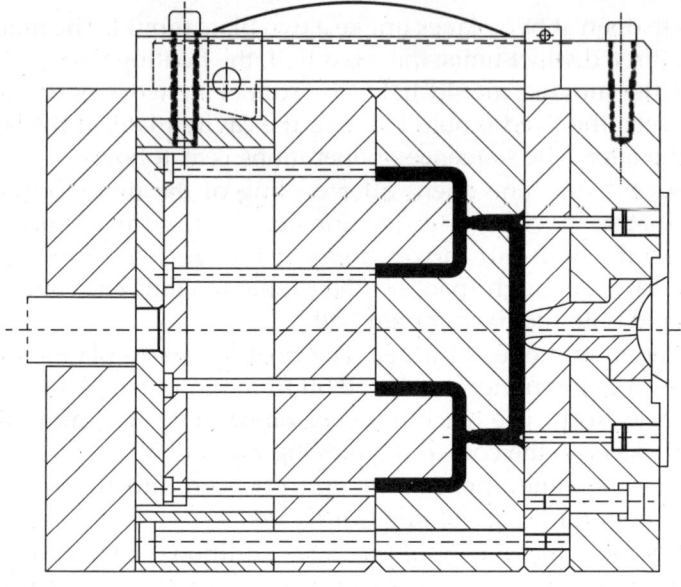

Fig. 4.70

the most suitable point and its automatic severing. For large products, any number of gates can be placed at desired locations.

The Self-insulating Runner Mould

Plastics are bad conductors of heat. If the runner were sufficiently thick, the plastics in the middle of it, insulated by the frozen outer layer, would remain hot and molten for a long time. Successive shots could be pumped through this insulated "pipe", without taking out the runner every time. The mould would have characteristics of a three plate mould without its drawbacks.

The system is simple and cheap and works very well if the cycle is short and regular. Figure 4.71 shows a self-insulating runner design.

The system is successfully employed with thermoplastics having a broad melting range such as polyethylene, polypropylene, polystyrene, etc. In case of an interruption in the cycle, a shut off and restart, the frozen runner must be taken out and the cycle restarted. The system, though uncomplicated and economical, may pose problems if the power supply is erratic.

The key point of the mould is the runner, its shape, thickness and length. The runner should be as short as the placement of cavities allows. A circular cross-section offers advantages over other forms. A minimum runner dia-meter of about 20 mm. has

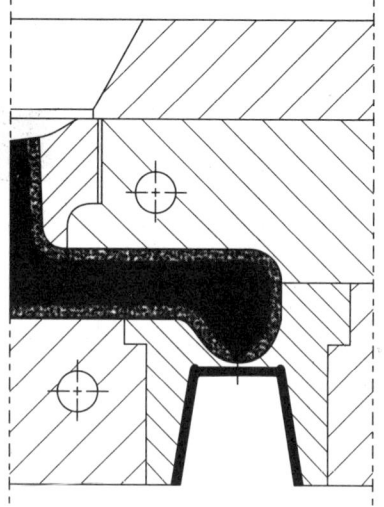

Fig. 4.71

been found necessary for moulding of polyolefines. One should keep some margin in the design for enlargement of runners after the try-out. The sprue should be short, well tapered and equal to the runner in size at the junction. The secondary runners too must be short, round and as big as the runner. The plates enclosing the runner should house the network for circulation of hot water/oil. The mould should be maintained at the highest temperature level permitted for the particular plastics. It may even be heated above 100°C to start with and, after a number of successful cycles, brought down gradually to the recommended temperature. The runner achieves an equilibrium if the molten core is renewed with each shot.

As the runner may cool down and may have to be removed after interruptions, an arrangement to open the mould quickly at that level, right on the machine, is a must. The arrangement employed in the depicted design (Fig. 4.72) consists of removing the bolt attaching the side clamp to the fixed half, swinging the latch and attaching it to the clamp bolt. Now the cavity plate becomes a part of the moving half. As the mould opens now, the cavity plate travels with the ejection side and the runner becomes accessible for removal.

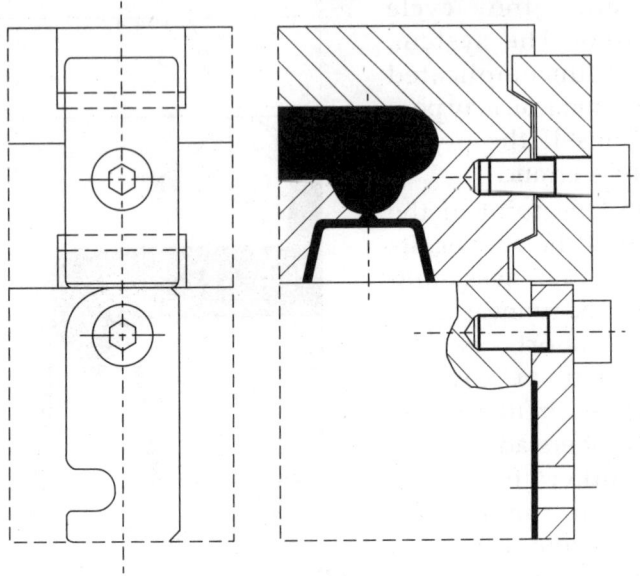

Fig. 4.72

The mould plates on the injection side should not be very thick so that the sprue remains short. These should be manufactured out of prehardened materials like P20. It must be noted that a thick runner exercises a large force which may deform plates made of ordinary materials.

The most critical point in the feeding system is the end of the secondary runner where the gate starts. Here, the frozen layer may become too thick to be pierced open with the injection pressure. Incorporation of short torpedoes, heated internally with slender cartridge heaters, can help keep the gates open (Fig. 4.73).

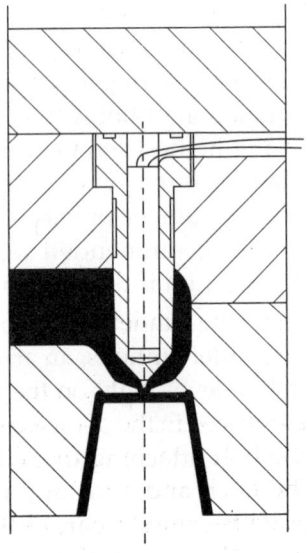

Fig. 4.73

The Hot Runner Mould

Another way of keeping runners in the molten condition is to provide a heating arrangement for them in the mould. It is analogous to extending the machine nozzle right up to the mould cavity. As the mould must be cooled to dissipate the heat contained in the moulding, one must segregate the runners from the mould body in order to preserve their heat. To effect this, the feeding system is constructed as a separate unit, consisting of a sprue bush, a manifold containing runners and the nozzles attached to the manifold bringing the melt to the individual cavities (Fig. 4.74). The unit is maintained at a temperature suitable for the flow of the melt. Though the unit is housed in the injection half of the mould, it is segregated from it as much as possible.

The Manifold

The primary runner and its branches are housed in a block called the manifold. It is usually a solid hot die steel block with drilled holes for runners and sub runners. The ends of the drilled holes are blocked with plugs which divert the melt to sub runners. The diverting pins also eliminate dead ends where the melt could stagnate and degenerate. The drilled holes for the melt forming the runners must be as smooth as possible to cause very little pressure loss through friction. Rough surface can also arrest polymer particles which could get released later

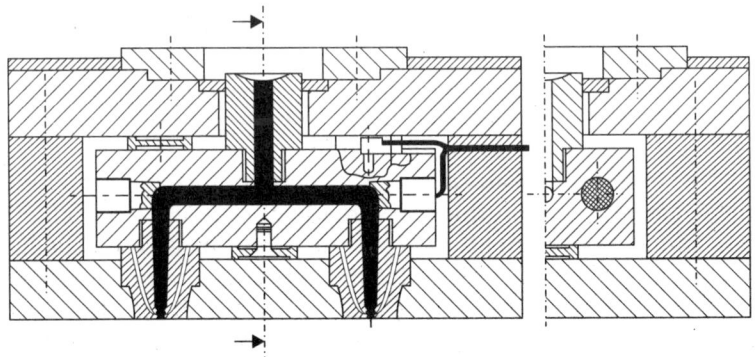

Fig. 4.74

and lead to rejection when plastics of another colour is moulded.

The size of the runners is chosen to restrict the dwell time of the melt in the runner system to one or two shots. However, too small a runner may result in considerable loss in pressure and overheating of the melt. Runner diameters less than 5 mm. are generally avoided. Too large a diameter, on the other hand, takes too long to melt when the cold mould is restarted.

The manifold block is heated by means of electrical heaters. The factors for determining the required heating capacity is the weight of the block and the heat conductivity of its material. As per a thumb rule, 200–300 watts are needed per kilogramme of steel. The source of heat may be cartridge heaters, inserted in drilled holes parallel to the runner channels. The holes must be accurate and well reamed so that the heaters sit snugly and transfer their heat efficiently. Air gaps may lead to accumulation of heat in the cartridge at untouching spots and to its burning out. The holes for the heaters should be through going to facilitate easy removal for replacement.

The cartridge heaters are built out of cylindrical metal tubes and contain heating spirals insulated by heat conductive magnesium oxide. They are cheap, easy to replace and are available in various sizes and surface watt densities. As they can be placed only in straight lines, they cannot heat all secondary runners equally and uniformly in case of multi-cavity moulds. Thermocouples for controlling and regulating the temperature may also be integrated in the heaters. It has, however, been found more convenient to place separate thermocouples in the manifold close to the injection points.

A point worth remembering while designing with heating cartridges is that these do not have heating spirals close to both ends for a length of about 10–15 mm.

The manifold is supported at crucial points by hardened discs, about 5 mm. in thickness, made of stainless steel which has much poorer heat conductivity than most tool steels. Titanium can also be set in to this end. The area of contact is further reduced by providing grooves on the discs. The discs also provide an insulating air gap all around the manifold. The heat losses due to convection are considerably reduced by sealing the gap to prevent air circulation.

The Sprue Bush

The sprue bush is a separate unit and is mostly screwed onto the manifold block. In simple moulds for commodity plastics, it need not be heated separately as it gets heat from the machine nozzle on one end and from the heated manifold on the other end.

The sprue bush is a round component, turned out of hot die steel and hardened. It has a straight hole along the axis for conveying the melt from the machine nozzle to the primary runner in the manifold. It has a spherical depression matching with the radius of the machine nozzle.

The sprue bush can also be provided with a sleeve like or a coil type heater with a thermocouple.

The Nozzles

The melt is conveyed from the manifold to individual cavities by the nozzles. There are two basic versions of nozzles, viz. the indirectly heated and the directly heated ones.

The indirectly heated nozzles are fabricated out of heat conducting metals like beryllium copper for self-fabricated hot runner units. These nozzles without own heaters get their heat from the manifold. Their exact temperature cannot be measured and regulated directly. An indirectly heated nozzle, as per BASF design, depicted in Fig. 4.75, heats the melt from outside. Another version of an indirectly heated nozzle or torpedo from Hoechst of Germany (Fig. 4.76) heats plastics from inside. The indirectly heated Be-Cu nozzle is surrounded by a hardened steel bush foreseen with an air jacket. In order to reduce the heat losses, the contact surface between the hot and cold components is reduced to the minimum. The manifold is sealed in the

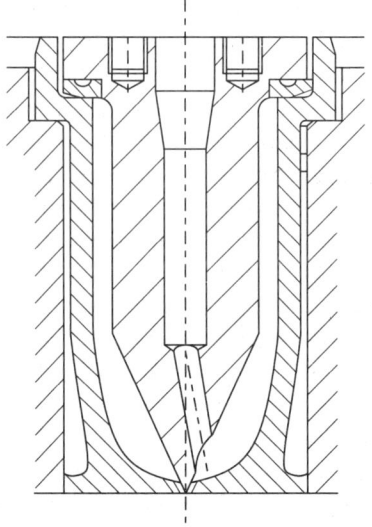

Fig. 4.75

housing to prevent heat losses by convection through air circulation. The system is simple, economical, space saving and feasible with easily available materials and components. It must, however, be pointed out that in this system, it is not possible to control and regulate the temperature of individual nozzles.

Indirectly heated nozzles work fairly satisfactorily with commodity plastics, which have a broad temperature range.

A directly heated torpedo like nozzle, which can be fabricated by the users, is shown in Fig. 4.77. It is heated by means of a cartridge heater inserted in a bore along the longitudinal axis. A thermocouple is built in the cartridge. The front part of the torpedo is foreseen with a tapering tip which keeps the gate open. The heated torpedo passes through the manifold and sits against the mould back plate. The melt from the manifold flows around the torpedo to the mould cavity. A major disadvantage of the commercial cartridge heaters is the absence of heating coil in the front part where the torpedo requires more heat to

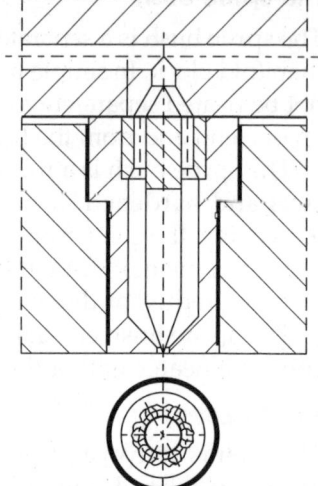

Fig. 4.76

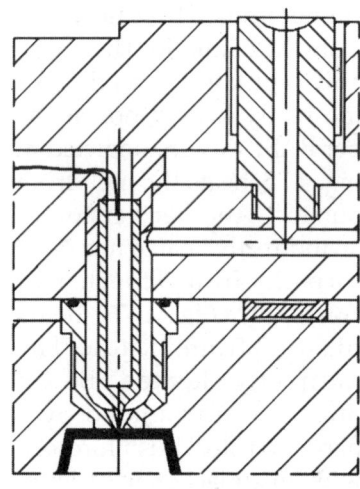

Fig. 4.77

keep the melt fluid and the gate open. The drawback can be compensated to some extent by making the body of the torpedo out of beryllium copper and the back part, which is held by the manifold and sits against the cold mould plate, out of hardenable stainless steel as shown here. It is worth remembering that the temperature measured and displayed by the thermocouple is that of the outer mantle of the cartridge heater.

Figure 4.78 illustrates a patented cartridge heater which is free of the shortcomings mentioned above. The heating is effective all over the length and the thermo-couple, which extends into the tip of the nozzle, measures the actual temperature.

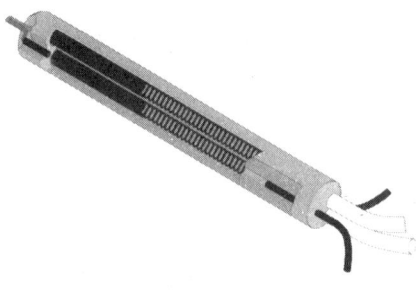

Fig. 4.78

Another recent deve-lopment comprises a hard pipe equipped with a jacket enclosing heating coils together with a thermocouple (Fig. 4.79). The front part of the pipe, jutting out of the heater, can be shaped as per requirement, e.g. a hard tip can be screwed in to form a hot runner nozzle. The head too can be

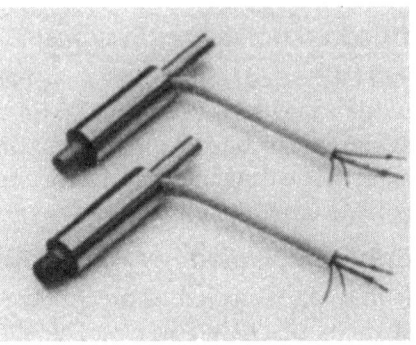

Fig. 4.79

machined to suit the self fabricated hot runner manifold for special cases.

The fixed mould half, housing the hot runner system, is bound to get heated because of the heat transferred. It needs very efficient cooling otherwise the linear expansion may make the mould closing difficult. Such moulds are generally foreseen with a special set of guiding devices, which align the mould halves without interfering with expansion, The device depicted here (Fig. 4.80) consists of a u-shaped female part and a flat finger fitting in it, is mounted outside on horizontal and vertical

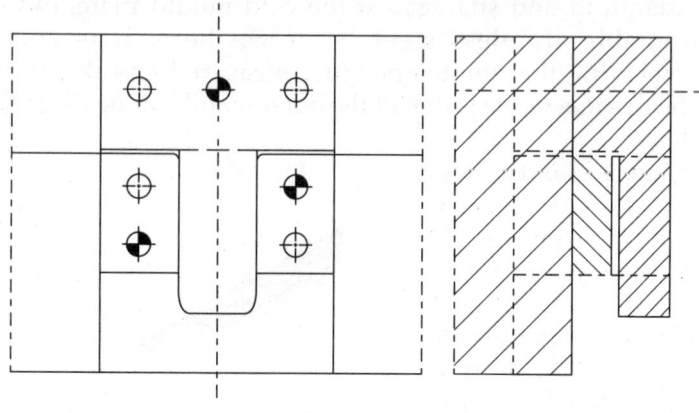

Fig. 4.80

axes of the mould plates. The guide pillars are given more clearance to eliminate fouling due to expansion.

Standard Hot Runner Systems

Self fabricated hot runners have been replaced, to a great extent, by standardised systems with direct heating. Most standardised systems are based on two distinct heating principles, viz. external heating and internal heating. The source of heat energy is invariably the electricity.

Externally Heated Hot Runner System

It may be regarded as a refined version of the self fabricated systems discussed so far. The essential components, viz. the manifold, the sprue bush and individual nozzles of the foregoing system play here too the same role. The difference lies in improvement of heating and standardisation. Figure 4.81 shows a typical externally heated system with usual components viz. the sprue bush, the manifold and the nozzles.

1. Register ring	2. Sprue bush
3. Support ring	4. Hot runner manifold
5. Multipole connector	6. Thermocouple
7. Nozzle cover	8. Heating coil
9. Nozzle body	10. Bush
11. Plastics melt	12. Hot runner nozzle
13. Cooling lines	14. Moulding

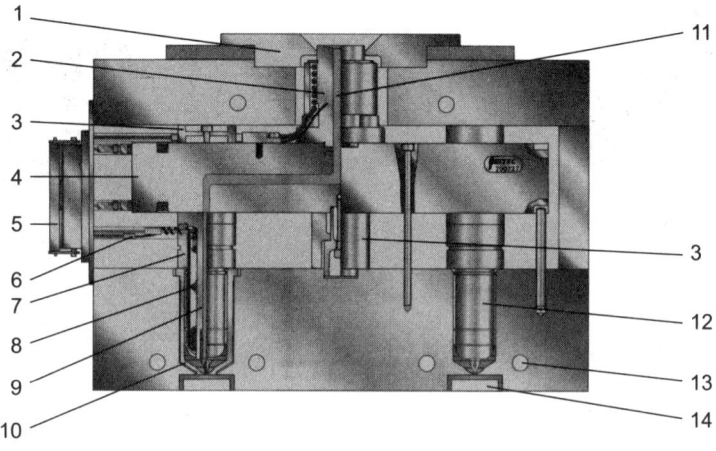

Fig. 4.81

The sprue bush, a hardened round component is either screwed onto the manifold or joined to it in some other way. It is generally fitted with a sleeve heater with individual thermocouple. It has a spherical depression on the face for seating the machine nozzle.

The manifold, like its predecessor, is generally a one-piece melt distribution block, equipped with controllable heating arrangement. Cartridge heaters have been replaced by tubular heaters for heating of the manifold. The tubular heaters can be bent and housed in channels machined on both sides of the manifold block to heat the runners uniformly. To intensify the heat transfer, channels are filled with a high temperature cement containing powdered copper.

The shape of the manifold is decided by the number of nozzles being fed by it. Standard manifold shapes are rectangle, H, X, circle or double X joined together (Fig. 4.82). However unusual shapes for special cases, as shown in Fig. 4.83, can also be fabricated and heated uniformly.

The inlaid heaters with conductive cement offer some significant advantages over the cartridge heaters:
- Larger area of heat transfer
- More uniform temperature distribution
- Freedom in location and number of cavities as highlighted in Fig. 4.83

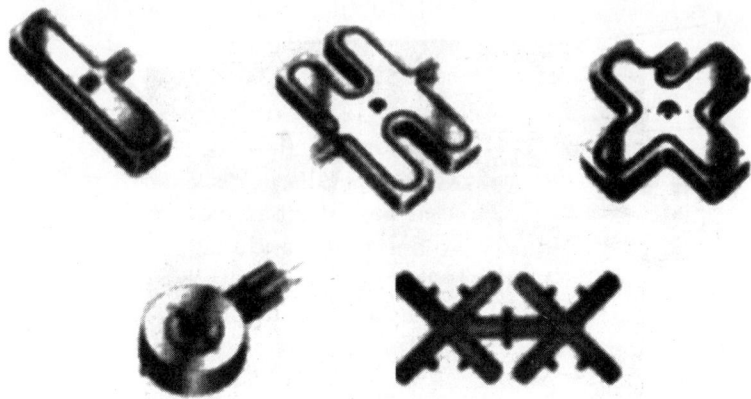

Fig. 4.82

A serious drawback of the method is the difficult replaceability of the profiled heater as it is fixed by means of a cement. It is more expensive and requires precise milling on both faces of the manifold. The bending operation of the tube requires special fixtures.

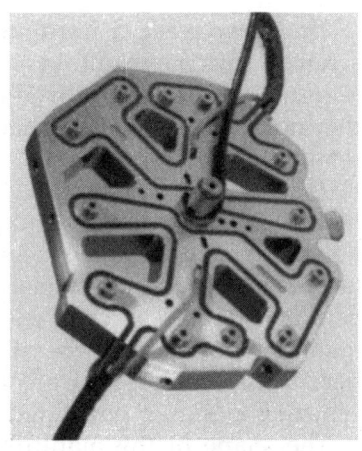

Fig. 4.83

A solution to the problem of replaceability, as practiced by one manufacturer of hot runners, is to add plates on both faces of the manifold and mill the channel for the profiled tubular heater partly on the manifold and partly in the covering plate. Insulating plates are added onto the covering plates and bolted to the manifold. Although the assembled manifold is more expensive, the easy replaceability of the heaters and considerable saving in heat energy make it economical in the long run. As the mould does not have to dissipate the heat convected and radiated to it from the manifold faces, the cooling time too is shorter.

A flexible version of the tubular heater with rectangular/ square cross-section can be placed in the milled channel manually without pre-bending.

An inevitable disadvantage of the drilled feeding network in solid manifolds is the sharp junctions of the runners and sub-runners which invariably results in loss of pressure. The latest method of diffusion welding has made it possible to divide the block into two half plates and machine the runners with rounded bends in both halves (Fig. 4.84). The surface finish of the channels can be optimised to facilitate easy flow and eliminate pit marks where material particles could stagnate. A perfect matching of contours in both halves is, of course, imperative. The two halves form a single solid plate after welding.

The manifold must sit snugly in the injection half of the mould to prevent leakage of the melt but contact with the relatively colder mould plates will result in undesirable transfer of heat to the mould. Here too, hardened discs made of a material with adequate toughness and poor heat conductivity such as stainless steel, titanium or even special ceramics are employed to support and fix the manifold between the clamping plate and the cavity plate. The supporting discs also create an insulating air gap between the manifold and the mould plates. The discs are located opposite the sprue bush

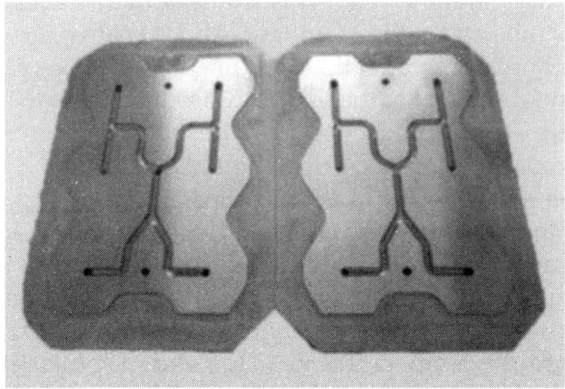

Fig. 4.84

and behind the nozzles leading the melt to the cavities to counter the bending forces on the manifold exerted by the machine nozzle on one hand and the injection pressure reaction on the other. Any deflection of the manifold caused by these forces will result in leakage of the melt, incomplete filling of cavities, filling of the insulating gap around the manifold, melting of the wiring of heating elements and finally to short circuits, destruction of heaters and total failure of the system.

Heat losses through radiation are much less as compared to those through conduction and convection and usually no special measures are undertaken to eliminate or reduce them. It is, however, possible to minimise the losses by covering the faces of the manifold block with thin reflector sheets made of aluminium alloy.

The nozzle. It represents the final, vital link between the melt feeding system and the cavity being filled. Its design should ensure supply of the melt with as little loss of pressure and heat as possible. The temperature too must be adjustable and controllable within very narrow limits.

In its basic form, an externally heated nozzle (Fig. 4.85) is a round, cylindrical bush with a slender body and a bigger head which butts against the manifold. Along its axis, it has a bore, the diameter of which equals that of the manifold at the junction but narrows down to form a gate at the exit end.

The open tip of the nozzle, as shown in the diagram, conveys the melt in a straight path. This configuration leaves a small vestige of the gate on the moulding. An alternative design (Fig. 4.85b) ending in a pointed tip extending into the gate, closes the melt channel and makes the melt flow around the tip to the gate. The hot tip keeps the gate open and the vestige is very small.

The nozzle is equipped with a coil heater around the body. It may be a sleeve (Fig. 4.86) or a heater coil as shown. The later has the advantage that the windings can be spaced closer or wider apart locally according to the amount of heat needed. For example, windings are closer near the exit point of the melt. A thermocouple is housed in a narrow slot on the nozzle body to measure the temperature near the end. A protective steel bush encloses the heated nozzle. The gate to the mould cavity is located in the bush which also forms a part of the cavity.

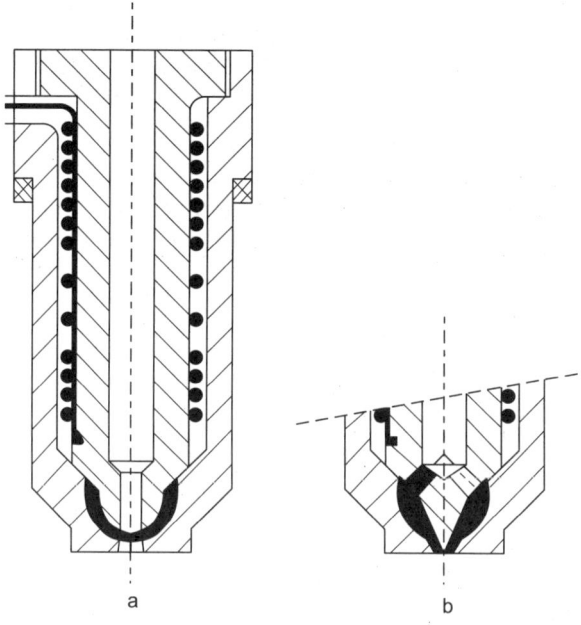

Fig. 4.85a and b

As obvious, the nozzle with heaters and protective bushes add to the size of the system. It proves disadvantageous in case of small articles, which cannot be located as close as their size would enable, resulting in less cavities in the mould. Multi-gating hot runner nozzles provide the solution. The cavities can be gated from the top (Figs 4.87 and 4.88) or from the side (Fig. 4.89). The temperature control is common for all cavities.

The nozzles described so far may be summed up as open type nozzles.

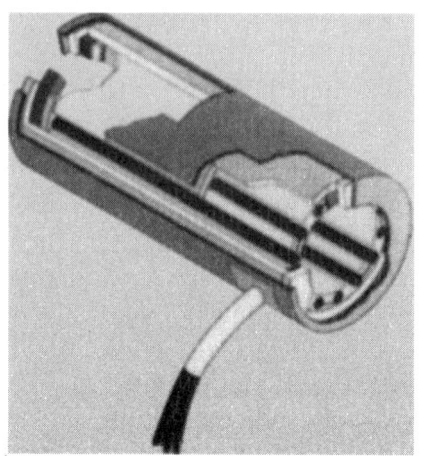

Fig. 4.86

A small gate vestige is inevitable. Secondly, these cannot prevent drooling of easy flowing polymers which require positive shut-off nozzles for trouble free moulding. It involves opening of the gate before injection and shutting it off at conclusion of the holding period. It is also termed as valve gating.

Fig. 4.87

One of the valve gating nozzles (Fig. 4.90) works on the same principle as the spring loaded shut-off machine nozzle. A needle with conical tip, placed in the nozzle in line with the gate and loaded with springs at the back, is pushed back with injection pressure, thus opening the gate. The compressed spring regains its initial height and pushes the valve needle forward to close the gate after the holding pressure is switched off. The device works satisfactorily if the holding pressure is not too low.

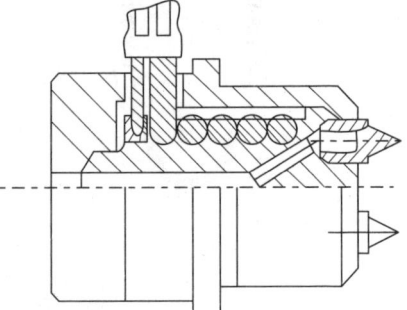

Fig. 4.88

Fig. 4.89

A more positive arrangement replaces springs with small pneumatic/hydraulic cylinders, built in the back plate of the mould (Fig. 4.91). The injection moulding machine must have controls for air/a core pulling arrangement

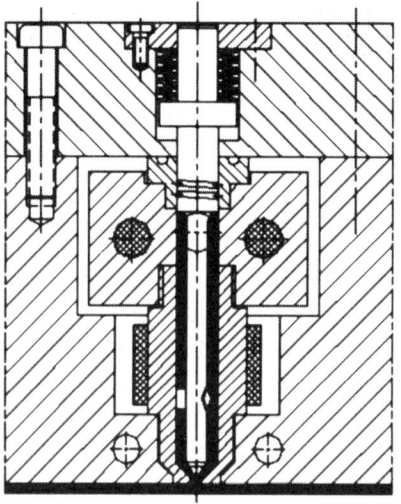 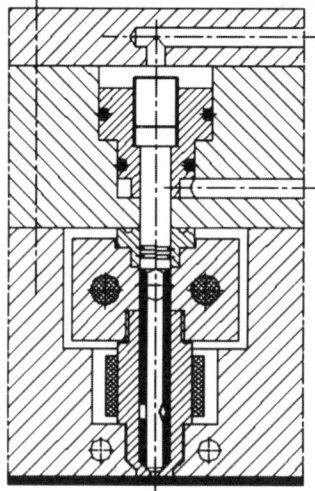

Fig. 4.90   **Fig. 4.91**

for their operation. One common cylinder can operate a number of needle valves in case of multi-gating nozzles.

Externally heated systems are distinguished by following advantages:

- Changing over to materials with different colour or altogether to a different material is unproblematic. Runners become free of the previous material after a few shots.
- The melt temperature can be measured and controlled accurately.
- Incorporation of valve gating is possible.

The system has also some disadvantages:

- Some heat is invariably lost to the mould housing. Effective thermal insulation between the mould clamping plate and the machine platen is indispensable to prevent conduction of heat.
- The manifold block undergoes linear expansion which changes centre distances of nozzles. The overall expansion may also cause melt leakage.
- Energy is consumed, not only to heat the plastics but also the manifold.

Internally Heated Hot Runner System

As the name implies, the material is heated from inside, in this system. The heaters are situated right in the stream of the melt. Figure 4.92 shows an arrangement with internally placed rods made of special alloys, which get heated when low voltage, high ampere current is passed through them.

The heaters situated in the middle of the runner in the manifold melt plastics around them. Here again, the poor heat conductivity of plastics prevents melting of the entire shell surrounding the heater. A cross-section of the runner would show plastics layers at various temperatures. The layer adjacent to the outer wall of the bore remains unmolten, whereas that in contact with the heater is fluid and at the desired temperature. There is a temperature gradient from centre to the periphery.

As the manifold remains cold, it need not be segregated from the mould body. There is no linear expansion either. The leakage too is eliminated by the frozen layer on all junctions. The heater in the nozzle keeps the gate open.

The Advantages of the Internal Heating System are:

- The plastics material is heated directly. The energy consumption is very low.
- The heaters heat only the core of the melt which flows.
- The outer layer of the melt in contact with the cold mould walls acts as a thermal insulator. The heat loss from the melt to the mould is very low. Insulating plates are not required.

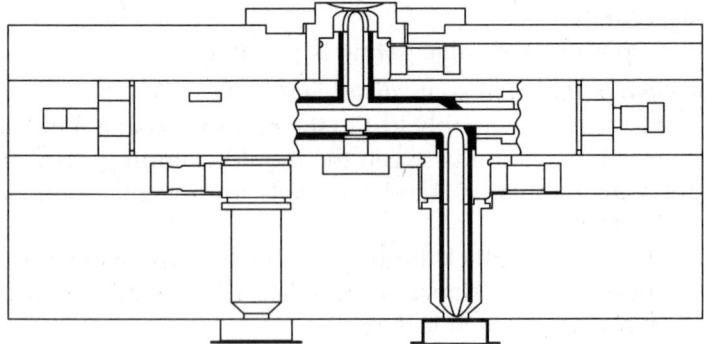

Fig. 4.92

- The manifold remains cold and does not undergo a linear expansion. No correction is required to be made for the change in position of the nozzles.
- The frozen outer layer prevents leakage of the melt.
- Runners and heaters may be incorporated right in the cavity plate. The mould becomes simpler and smaller.
- Because of the low voltage, the system is electrically harmless. It does not require special electrical insulation.

Some Drawbacks of the System are:

- The heaters are a source of additional friction to the melt flow. More pressure is needed for injection.
- Between the frozen outer layer and the fluid core of the melt, there are semi-frozen layers from which cold particles may be dragged along by the melt into the cavity.
- It is difficult to measure the temperature of the melt.
- It is not possible to incorporate valve gating through the heaters.
- Change of material or colour is difficult. Some particles from the previous material, frozen along the surface of the bore may occasionally get into the melt stream and cause rejection.
- The low voltage requires step down transformers.

Some hybrid systems combine both types of heating to make the best of both principles.

Hot Runner with Cascade Injection

Large articles may need more than one injection points for complete filling. This may be accompanied by weld lines and even air entrapment at the junction of different melt streams. Cascade injection, viz. sequential opening of the nozzles can eliminate the defect. The injection starts at one end and the next gate is opened when the melt from the foregoing gate has reached it. Likewise, consecutive gates are opened in sequence. The nozzles are, of course, equipped with valve gating and are regulated through a special controlling system, synchronised with machine controls.

The Stack Mould

Injection moulds for articles with large projected area but comparatively less weight such as shallow plates, lids, covers, trays, audio and video cassette bodies, etc. utilise only a fraction of the plasticising capacity of a given moulding machine because a limited number of impressions can be accommodated in the mould, both from the angle of the locking force as well as due to large space requirement. A 2-stack mould design enables accommodation of twice as many cavities in the mould as a normal mould and can be operated on the same machine without calling for additional locking force.

A stack mould may be seen as a combination of two identical conventional moulds, placed back to back with a common melt distribution system housed between them (Fig. 4.93). The feed system is usually of the hot runner type although theoretically a cold runner can also be employed. Apart from the feed system, the two moulds possess the usual components and design features. An additional device, characteristic of stack moulds, is the mechanism to open the mould at two levels in an equidistant way. The system illustrated in the present example employs a central pinion and two racks fixed to the two ejection mould halves. The same results can also be achieved with toggle levers, pivoted on the central plate and attached to the "fixed" mould halves.

The hot runner feed system may be self-fabricated as shown here. The manifold is made of hot working steels like 1.2344 and provided with cartridge heaters to make up for the heat lost by conduction, convection and radiation. The conduction losses can be kept to the minimum by reducing the area of contact between the hot manifold and the colder mould plates and the convection can be minimised by enclosing the heated manifold from all sides. A thermocouple helps keep the temperature within the pre-set limits. The nozzles, which convey the melt from the manifold to the runners, are foreseen with their individual heater coils and controlled through thermocouples. The exceptional feature of the system is a long, heated sprue bush, going through the right hand mould and projecting out of it.

Standard hot runner units, with special manifolds for stack moulds, can also be employed instead of the one described.

The ejection plates on the left are moved in conventional manner through the ejection mechanism of the machine but those in the right-hand mould need an additional device. These may be pullers as employed in the present case or toggle levers which move the ejector plates when the mould opens. Hydraulic cylinders too can be incorporated in the mould to perform this function.

A stack mould, obviously takes up more daylight in the machine. However, the shallow articles, generally moulded with stack moulds, do not call for large openings. Mostly, the normal injection moulding machines can be employed for stack moulds. Only in rare cases, it may be found necessary to set in machines with enlarged mould height capacity. However, the locking force remains unaffected as the injection pressures acting on two levels neutralise each other inwardly so that the net locking force remains the same as for a single mould.

A stack mould with two layers of cavities increases the production almost two-folds.

It is also possible to have more than two stacks, with corresponding changes in feeding and ejection systems.

The Temperature Regulation of Hot Runner Systems

Efficiency of a hot runner system is coupled with the precision and constancy of temperatures of its various components. Special controllers have been developed to this end. Most controllers consist of individual modules for each circuit and work according to the PID controlling characteristics. The sophisticated ones are generally equipped with microprocessors.

They Perform Following Functions:

- Control the set temperatures and maintain them in close limits.
- Automatic soft start; heating up with lower power
- Automatic switch over to calculated average power supply in case of thermocouple damage
- Have a leakage current control
- Protection against overheating of the mould
- Have a memory function
- Possibility of remote monitoring
- Fault diagnosis, display and protocol functions
- Give visual and/ or audio alarm in case of failure.

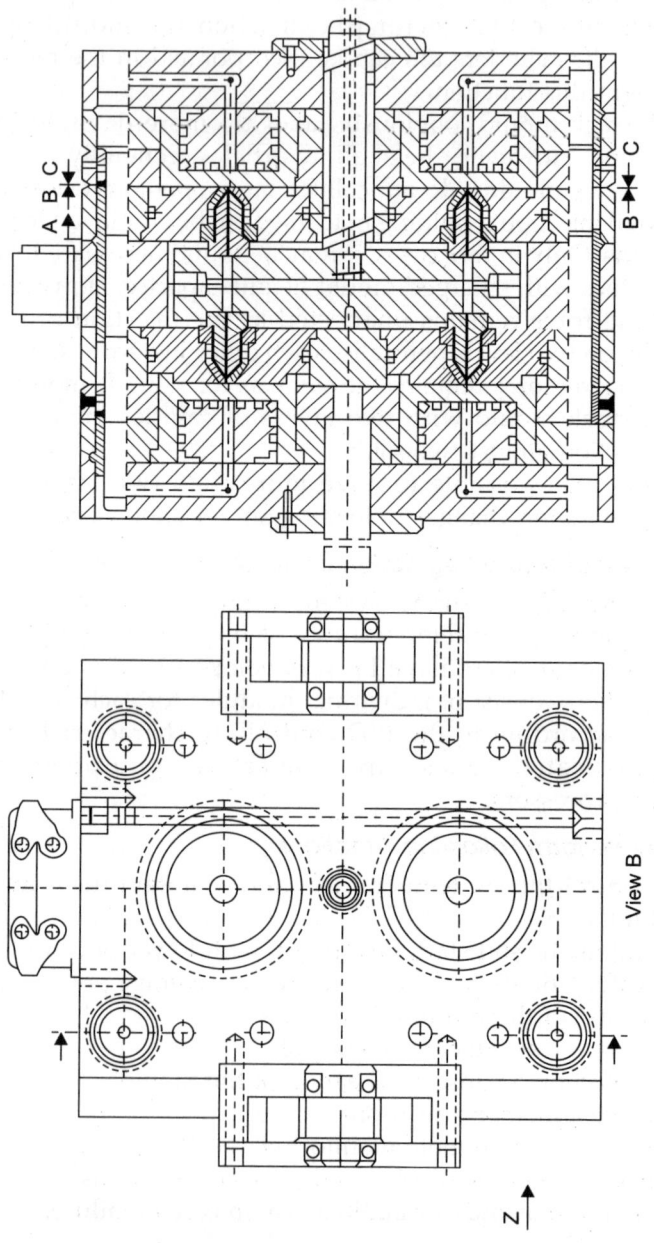

View B

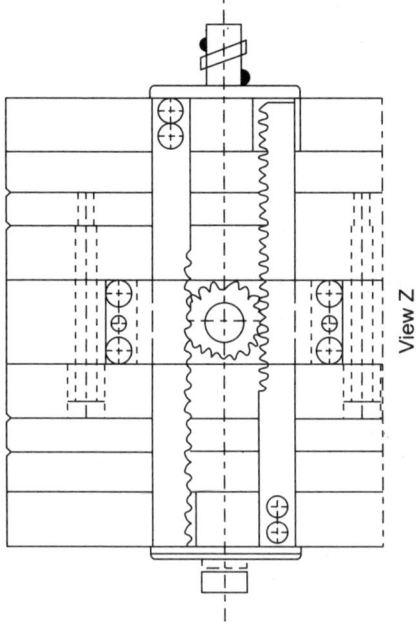

View Z

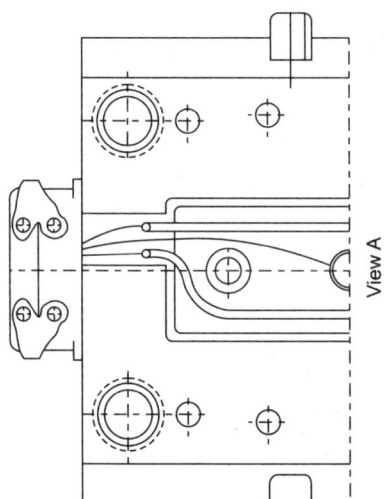

View A

Fig. 4.93

- Report deviations and faults.
- Possibility of coupling with the moulding machine controls
- Automatic temperature reduction during production interruption /disturbance
- Have interface for printer

One regulating module can control an individual circuit or a number of circuits joined together in case of large multi-cavity moulds. In the later case, all circuits governed by one module will be treated as identical. For example, 16 modules could regulate the temperature of 64 nozzles of a mould which are wired in groups of four.

The latest version of temperature controllers consists of versatile modules which are capable of regulating a number of circuits individually in all respects.

Conclusions

The advantages of a hot runner system are:

- No runners have to be cooled, ejected, reground and remixed.
- The mould is simpler as it does not have to open (and close) at two levels.
- The temperature of the melt to all cavities can be maintained at the required level.
- The balancing of filling can be manipulated through variation of temperatures of the nozzles.
- The product can be gated at the most suitable points.
- The melt reaches the cavity at the predetermined temperature.
- The machine needs less energy as it has to melt, cool, and inject only the material for the cavities and not for the runners.
- The production cycle is shorter as no runners have to be injected, cooled and ejected.
- The machine can be smaller and hence cheaper.

Disadvantages of hot runner systems are:

- High initial investment in the mould.
- Need for additional regulating equipment
- More chances of disturbance due to failure of heaters and thermocouples

- Possibilities of leakage at junctions of nozzles and the manifold
- Delay at restart

The cumulative effect of above points is a possibility of automation, uniformity in quality, decrease in rejection and reduction in the cycle time.

When using glass reinforced resins with hot runner moulds, it is advisable to make the first shot with unfilled material. The resident material forming the insulating cap around the nozzle tip is less conducting than the filled material, which makes it easier while restarting after a pause.

Unscrewing Moulds

Unscrewing with Archimedean Screw

It is a method to derive a rotary movement from the linear movement of the mould. It employs a long pitch screw, fixed on the injection side of the mould, which rotates a nut housed in the moving half when the mould opens. The periphery of the nut is either formed like a gear or else a gear is fixed onto it. This acts as a sun gear and transfers the rotary motion to the planet gears fixed on the threaded cores of the mould cavities (Fig. 4.94).

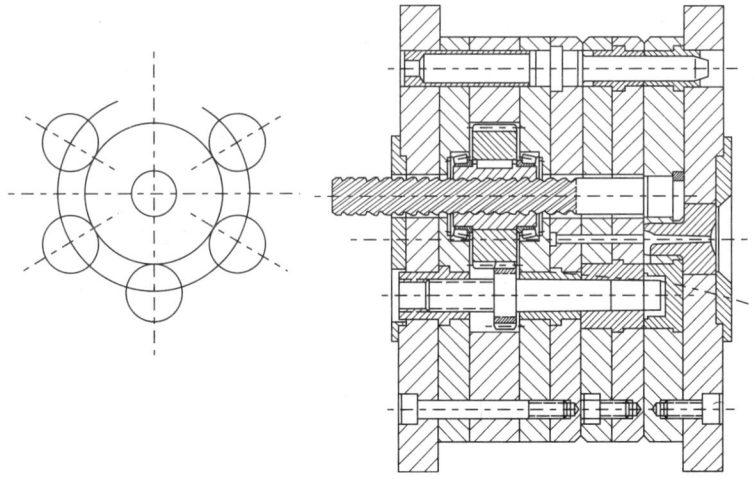

Fig. 4.94

The number of revolutions achieved by the central gear is decided by the pitch of the Archimedean screw and the opening stroke of the machine. These revolutions have to be converted into the number of revolutions a threaded core has to make to be free of the moulded threads in the product. In other words, The size of the sun and planet gears and consequently the distance of the cavities from the centre and the size of the mould, is dependent upon the opening stroke of the machine, the pitch of the screw and the number of threads to be unscrewed.

The pitch of the multi-start Archimedean screw is selected in relationship with its diameter. The aim is to keep the helix angle between 30 and 40 degrees so that the major part of the force acting between the screw and the nut (here the inside of the sun gear) goes towards rotation.

All rotating tool components must be provided with bearings. The bronze bushes are employed in moulds destined to produce small quantities. Ball bearings reduce friction and prolong the mould life. The bearings for the sun gear have to be of the combined type as these are required to guide axially and take up thrust too.

It may be added that an unscrewing mould with Archimedean screw can be designed in a number of variations. The screw may also be incorporated in the moving half of the mould and the nut may be fixed on the third platen behind the moving platen or else it may be pushed with the help of the ejection device. The principle of converting linear motion into a rotary one with the help of an Archimedean screw remains the same.

Unscrewing with Rack and Pinion

The linear movement of the machine platen can also be employed to obtain rotary motion needed for unscrewing, with the help of a rack (A) and a pinion (B) as illustrated in Fig. 4.95. Rack A, fixed to the stationary mould half, rotates the pinion B mounted on the moving mould half, which in turn operates another rack (C). The aim is to derive linear movement at right angles to the movement of the machine platen. Now the rack C can rotate a number of pinions attached to the threaded mould cores, housed in the moving half.

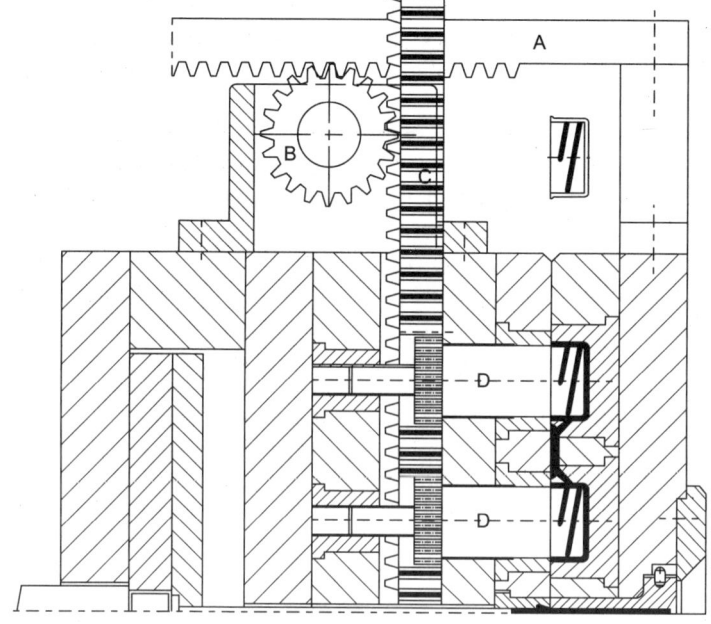

Fig. 4.95

The method, although also mechanical, differs from the foregoing one in that no intermediate sun gear is involved and the rack transfers the movement directly to the mould cores. These are provided with extensions having male threads of the same pitch as the product, reaching into threaded bushes as also in the previous example. Upon rotation, the cores are made to recede by the same amount as they disengage from the moulding so that the latter is free to fall down after complete unscrewing. The cores return to their initial position as the mould closes. Here too, the opening stroke of the machine imposes a limit to the number of revolutions achievable.

As obvious from the sketch, the rack A has to extend over and beyond the moving platen. The height of the main pinion B has to be chosen accordingly.

Unscrewing with Hydraulic Cylinder

The process of unscrewing can be made independent of the machine stroke if a hydraulic cylinder is employed as the primary source of linear movement of the rack C of the foregoing example, responsible for rotating the mould cores (Fig. 4.96). Obviously, rack A and pinion B are not needed in this variation.

Here too, the cavities are arranged in rows and straight racks, engaged with the gears on the threaded cores, rotate them, when they are imparted linear movement by a hydraulic cylinder attached at the end. The cylinder/cylinders can be placed above or below as convenient.

The process has an additional advantage that it can be carried out at any stage of the moulding cycle, also when the mould is still closed. The cores come to their initial position when the mould is closed for injection, as in the foregoing case.

A module between 1–2 mm. for the rack and the gears is found to be adequate in most cases. The larger number of teeth facilitates finer adjustment of threaded cores.

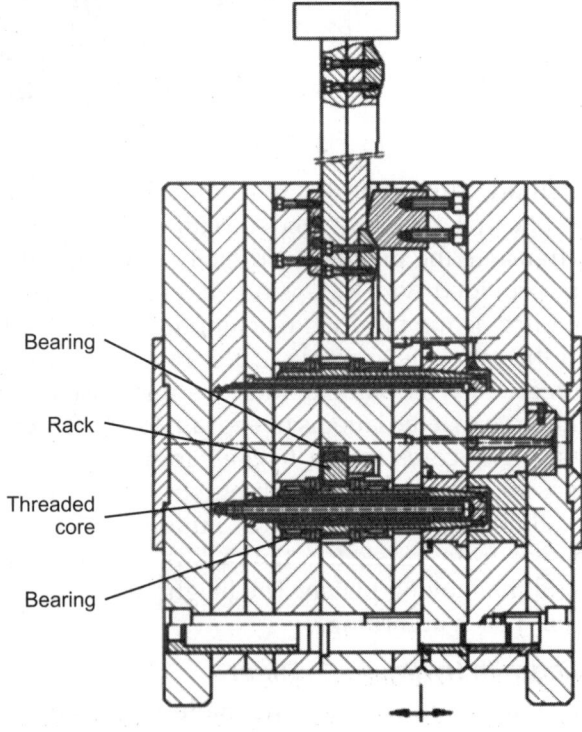

Bearing

Rack

Threaded core

Bearing

Fig. 4.96

Unscrewing with a Hydraulic/Electric Motor

The rotary movement required for unscrewing can also be obtained from a hydraulic or an electric motor. These can be mounted on the mould and coupled to the sun gear with intermediate gears, chains or belts. These devices are in no way dependent on the stroke of the machine and can provide unlimited number of revolutions. However, they rotate in one direction only and cannot be limited exactly to the revolutions corresponding to the number of threads in the moulding. The fact becomes significant if the start of the thread in the moulding has always to be in a particular position. This method cannot always guarantee the same position. Instead of making the core recede during unscrewing and advance to the original position before moulding as it happens with a rack and pinion arrangement, the cores rotate in place and the stripper plate is made to move forwards with the help of springs (Fig. 4.97).

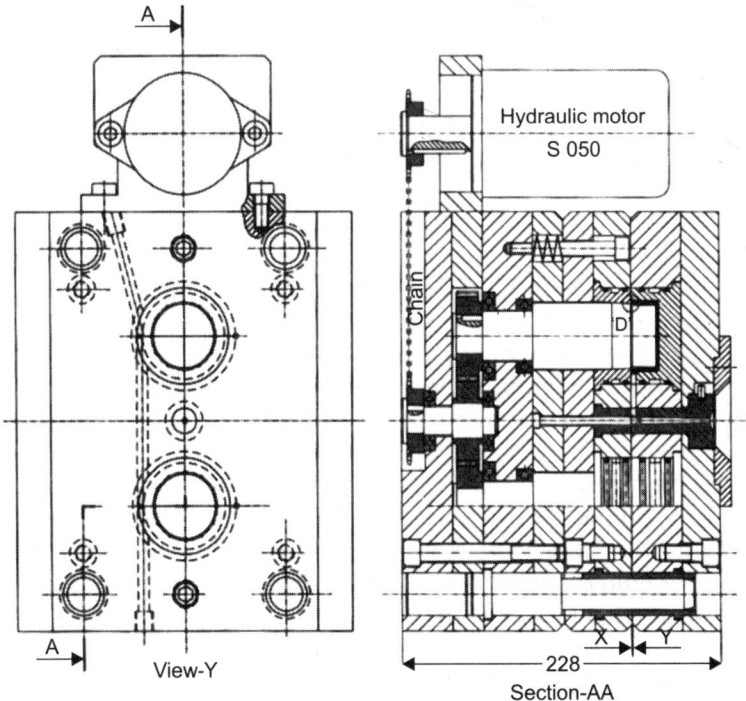

Fig. 4.97

The mouldings are retained on the stripper plate by means of the notches provided on their periphery till the unscrewing is almost complete. The stripper plate is arrested at this stage and by further unscrewing, the moulding moves further, becomes free from the holding notches and falls off.

The alternative method of rotating the cores without moving them linearly offers an additional advantage. It is possible to cool these components if they are not very slender, whereas cooling the cores performing both rotary and linear movements is very difficult.

To operate a hydraulic motor, one needs additional machine accessory called the core-pulling device, regulating delivery and exhaust of oil under pressure, in the moulding machine. In absence of that, hydraulic connections and controls from the ejector can be employed to operate the motor.

The method needs less number of components in the mould, as there is no conversion of a linear motion into a rotary one.

CLASSIFICATION OF INJECTION MOULDS

In order to describe the type, construction, complicacy, etc. through the nomenclature, injection moulds are classified according to various design features such as the number of cavities, number of parting lines, the feeding system, the ejection system, special features, etc. The broad categories are:

1. **Standard mould:** It is the simplest type of injection mould, consisting of two basic units; the cavity side (fixed half) and the core side (moving half). It is also referred to as a two-plate mould. The feeding is through cold runners and ejection through ejector pins and/ or sleeves. Figure 4.1 depicts a standard, two plate mould.

2. **Semi-automatic/ fully automatic mould:** Depending upon the purpose of production or the quantity needed, a mould may be very simple with a partial manual operation, say for putting in a core or an insert before injection. Most production moulds operate fully automatic.

3. **Single cavity/multi-cavity mould:** This indicates the number of impressions in a mould. The consideration for the choice may be the shape of the product, the size of

the machine, or the quantity required. The cost varies accordingly.

4. **Cold runner mould:** A mould with conventional feeding system which is cooled and ejected out along with the product is termed as a cold runner mould. The mould shown in Fig 4.1 is an example of a cold runner mould.

5. **Three plate mould:** As described before, the runner of a three plate mould is not on the same level as the moulding and necessitates another floating plate unit for ejection of the former. It has been illustrated in Fig. 4.70.

6. **Semi hot runner/ full hot runner mould:** Semi hot runner mould has a part of its feed lines in a heated block. The last channels leading to the cavities are in conventional form and are cooled and ejected.

   A full hot runner mould has no runners to be ejected.

7. **Stripper plate mould:** This classification is on the basis of the ejection system. Here a plate strips off the moulding or mouldings from the core/s.

8. **Split cavity mould:** The mould is named so when the outer contour of the article is formed by movable splits,

9. **Slider mould:** When some details in the article like side depressions (such as holes) or projections (such as plugs) prevent ejection and have to be cleared with the help of an external cross slide before the moulding can be ejected.

10. **Unscrewing mould:** Mould for an article with internal threads, incorporating a device to clear the threads before ejection.

Most moulds contain more than one of the above features and hence it is not possible to restrict their classification to one particular category. The descriptive nomenclature emphasizes the salient features, such as "Four cavity hot runner unscrewing mould" or a "Split cavity stripper plate mould".

PROCEDURE FOR DESIGNING AN INJECTION MOULD

1. Collect full information regarding:
   a. The product—mode of use; visually, functionally and dimensionally important features; matching and mating parts if any; after-operations; weight, colour and surface finish; quantity.

    b. The plastics—filled or unfilled; flow properties; moulding and mould temperature, shrinkage and its pattern; rigidity.

    c. Injection moulding machine—locking force and shot weight; platen drawing, opening stroke, height of the mould; nozzle type and its position, nozzle radius and bore diameter; ejection mechanism and the stroke; position of air and cooling water sources; ancillary equipment etc.

    d. The tool materials—steels specified and available; possibilities of hardening and polishing.

    e. Standard mould components—specified; available in stores and market.

    f. The quotation—terms concerning the tool.

2. Study the design of the product.

    Check the dimensions. Are they correct? Are all relevant dimensions given? Are tolerances specified? Are they feasible?

    Examine the design for special features like undercuts, threads, holes, metal inserts, etc.

    Will these make the mould design more complicated? Will a modification simplify the mould without jeopardising the function?

    Is the product design compatible with the specified plastics? Any unusual standards to be followed?

3. Decide the number of cavities.

    Are these specified? Are they limited by the machine capacity? Should these be decided on the basis of profitability?

4. Decide the mode of operation of the mould.

    Manual, semi-automatic or fully automatic. Is it specified?

5. Select the mould parting line.

6. Choose the position and type and number of the gates if not specified (It also influences the type of the mould, viz. 2-plate, 3-plate, hot runner). Check the melt flow, weld lines, air entrapment, sink marks etc.

7. Select the most suitable arrangement of cavities if not specified.

8. Ascertain the taper required and permitted.
   In absence of a specifically mentioned taper, split the tolerances to create taper.
   Check whether reverse taper is called for to retain the moulding on the ejection side.

9. Calculate the mould dimensions (product dimensions+ shrinkage allowance).

10. Decide the mode of placing and securing inserts to be moulded-in, if any.

11. Decide the mode, elements and location of ejection keeping in mind the visible areas.

12. Lay out the sectional elevation and the plan of the mould simultaneously. Build up the mould around the article. Draw the mould in the way, as it would be fitted on the machine.

13. Draw ejectors in the plan and then lay out the cooling, keeping in mind the hot spots. Make sure that the ejector pins, return pins and pullers etc. do not come in the way of free fall of mouldings and runners. The water inlet and outlets should be away from the hot runner connections and should not foul with clamping position of the mould on machine.

14. The length of the guide pillars should ensure that the two mould halves are in engagement before a mould core enters the cavity.

15. The assembly drawing should depict the function of the mould, the interrelationship of the mould components and their mode of assembly and matching.

16. Review, which components are more prone to damage or excessive wear and tear. Design them as easily replaceable inserts.

17. As far as possible, draw the individual components in detail as the machinist will see them and dimension them accordingly. Keep the mating parts close to one another to facilitate quick cross-checking.

18. Specify material and the treatment for all components. Indicate the surface finish and tolerances/ fits wherever applicable. Put short but clear remarks wherever the function, matching, fit, or accuracy cannot be expressed in figures.

19. Specify the cavity identification, its size and location. **Select the simplest one among various possible design solutions!**

DESIGN OF INJECTION MOULDED ARTICLES

The function of an article decides its shape and size as well as the most suitable material for it. In order to select the right material for a new component, one has to analyse all conditions—mechanical, electrical, chemical, thermal, optical and commercial—which the new article will be subjected to. The following list facilitates the selection.

1. The function
2. Environmental conditions; sunlight, UV rays, direct and radiated heat, humidity, sand and dust, fluids, fumes and gases, snow and ice, etc.
3. Temperature range; short term/long term
4. Forces; tensile, compressive, shear, bending, impact
5. Electrical requirements
6. Resistance to chemicals
7. Permeability to gases
8. Resistance to wear, scratch resistance
9. Fireproofness, inflammability
10. Conformance to food laws
11. Conformance to safety standards
12. Surface finish
13. Clarity, transparency
14. Colour assortment
15. Dimensional accuracy, tolerances
16. Matching components, type of matching
17. Method of assembly such as riveting, bolting, pasting, welding, etc.
18. After-operations; machining, printing, embossing, painting, metallising
19. Special features; integral hinge, snap fitting, multi component moulding, etc.
20. The present material, process of forming and design
21. Price range

Irrespective of the thermoplastics chosen, certain guidelines which ensure, not only a sound component but also an efficient, sturdy and economical mould and a trouble free moulding, must be followed while designing the article. These are:

Uniform Wall Thickness

This must be regarded as a universal rule, applicable under all conditions. It ensures uniform mould filling, uniform cooling, uniform shrinkage and therefore less stresses and less warpage. Thicker sections (Fig. 4.98A) are bound to cave in upon moulding as shown through dotted lines. The article can be modified as depicted in Fig. 4.98B to fulfil the same function without being prone to the said defect. Wall thickness is primarily decided upon from the strength angle. Another factor governing the wall thickness, no less important, is the possible distance the selected material can flow with the given wall thickness before freezing. It is expressed as L/T ratio.

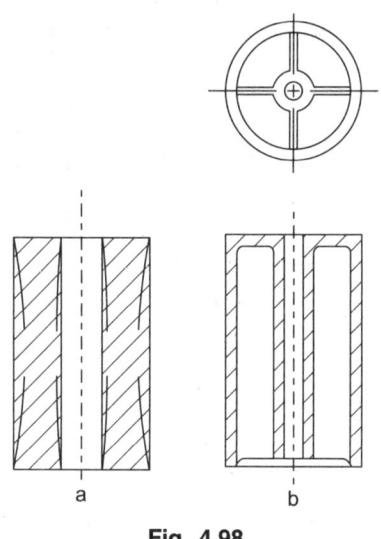

Fig. 4.98

Strengthening Ribs

Large surfaces can be strengthened by providing ribs instead of increasing the wall thickness. Thick sections cool slowly, prolong the cycle and may get sink marks. Ribs reduce material, decrease cooling time, add stiffness and strength and reduce warpage.

Thick ribs, too, can give rise to sink marks at their junction. Their breadth should not exceed two thirds of the thickness of the wall they join. For ribs which are joined to the article wall on one side only, the height is limited to thrice their breadth. A small fillet radius at the joint reduces the notch effect.

Cross ribs on a large area generate another source of moulding defects. There is unavoidable material accumulation where two ribs cross each other. One measure to reduce the material at the junction is to place a blind hole there (Fig. 4.99). It may, however, necessitate additional ejectors. The alternative remedy is to stagger the ribs as shown in Fig. 4.100.

Depending upon their direction, ribs may act as additional runners and give rise to separate melt streams leading to air entrapment.

Hexagonally arranged ribs lend greater rigidity than those in square array, without increasing weight.

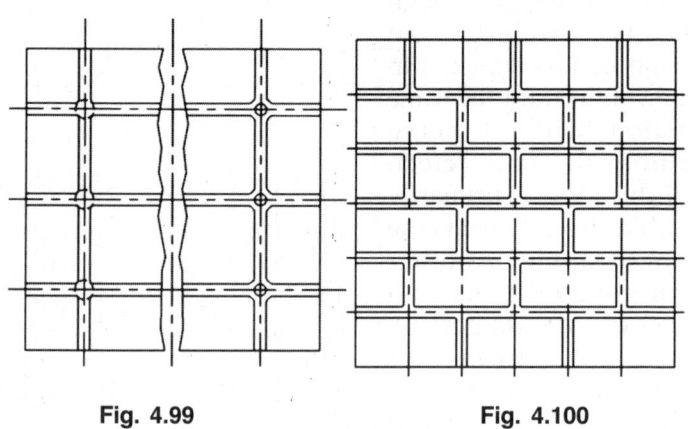

| Fig. 4.99 | Fig. 4.100 |

Gradual Variation of Wall Thickness

Thin sections cool down and shrink while the thicker ones are still in the process of cooling and shrinking. It leads to stresses and warpage. The changes in thickness, if unavoidable, should be gradual.

Relaxing Walls

Straight walls of box like articles tend to cave in, especially when moulded in semi-crystalline thermoplastics. A light convexity in design can counter this tendency. The walls tend to become straight after cooling and shrinkage.

Rim Reinforcement

Hollow containers like buckets, boxes and bins become stiffer if provided with reinforcement on rims. However, a solid beading or extra thickening leads to sink marks, distortion and long cycles. A hollow beading, semi circular or rectangular in section lends additional strength without any of these defects (Fig. 4.101).

Base Design

Large symmetrical containers like boxes, lids, buckets, basins, tubs etc., injected in the centre of the base, tend to develop cracks there because of concentration of moulding stresses around the gate and due to differential shrinkage (different along the flow and across it) in semi-crystalline plastics. A lens shaped base (Fig. 4.102), especially in the case of round articles, helps release the stresses. Same effect can be achieved by resorting to a pyramid shaped base for rectangular articles (Fig. 4.103). Internal stresses flatten the dome or the pyramid and thus get released without causing damage to the moulding. A

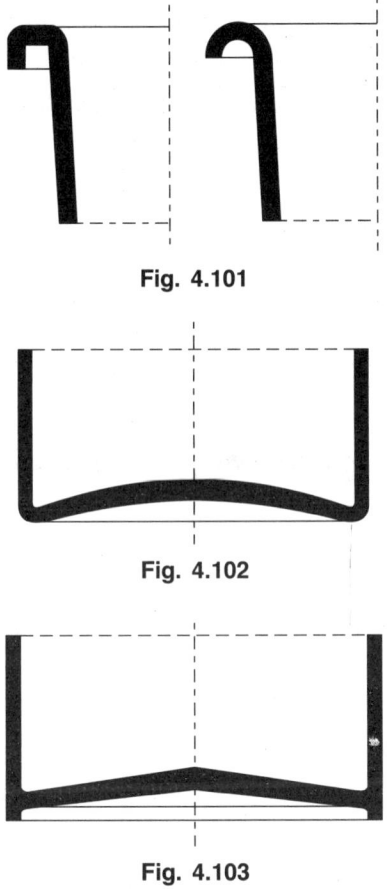

Fig. 4.101

Fig. 4.102

Fig. 4.103

base, slightly thicker in the middle and tapering down to the normal wall thickness towards the side walls offers the added advantage of aiding the melt flow with less pressure and providing strength at the point of maximum stresses.

Rounded Edges and Corners

Sharp corners lead to accumulation of material. Their cooling is much slower because of heat accumulation and because corners of mould cores cannot be cooled as efficiently as the outside surface of a moulding through the cavity plate/ insert. The result is non-uniform shrinkage and concentration of stresses causing distortion and cracks.

Taper

Taper in direction of ejection for distortion free demoulding. A minimum draft angle of 1 degree is recommendable.

Sharp Notches Facilitate Cracks

Sharp edges and corners, if unavoidable, must be rounded off. Provide fillet radii at the junctions of pins, bosses, ribs, etc.

Lettering

Raised lettering on a moulding, though easiest to make in the mould, is liable to damage and wear in use. Sunken letters are safe but difficult/more expensive to make. The combination viz. raised letters on a depressed surface offer the best of both (Fig. 4.104).

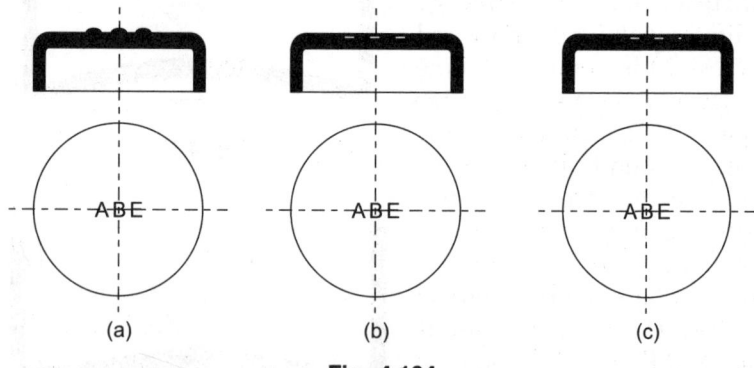

(a)   (b)   (c)

Fig. 4.104

Assembly Aids

Assembly aids in industrial components should be situated away from edges and corners. The holes for location of punches or cores in the mould forming the desired holes, when drilled too close to the edge of the mould cavity or core, weaken it considerably. Figure 4.105 illustrates the point and also the alternative.

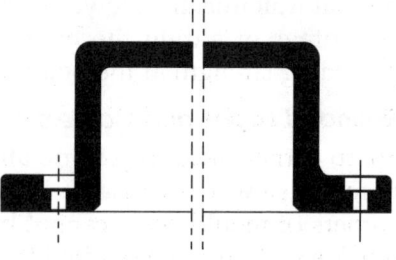

Fig. 4.105

Attachment Lugs

Attachment lugs for assembly of the component should be strong. Avoid sharp-edged changes in section as well as holes at the change of sections. Place attachment points outside, rather than inside (Fig. 4.106). Lugs located inside constitute undercuts and necessitate special devices like tapered slides in the mould, which increase its complexity and cost.

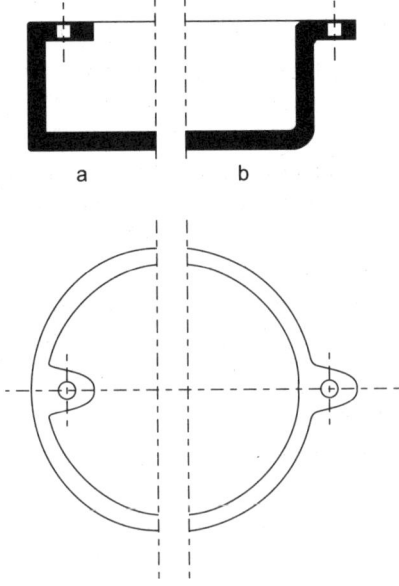

The Bosses

The bosses (Fig. 4.107) should be preferably hollow. Their wall thick-

Fig. 4.106

ness, taper and radii should follow the rules applicable to the ribs.

The boses should not be too close to or joined to the walls. For support, one or more ribs may be used as shown in Fig. 4.108.

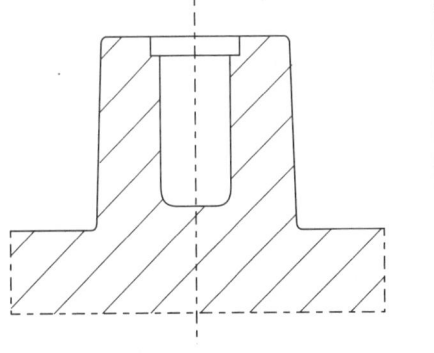

Fig. 4.107

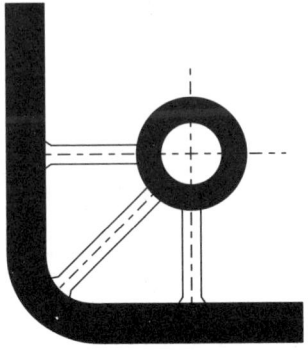

Fig. 4.108

Bosses with moulded holes for self tapping screws should be well tapered outside and should have fillet radii where they join the article. The hole for the screw should have a diameter ranging between 0.7 to 0.95 times the screw diameter. The smaller core hole diameter is applicable to softer plastics and the larger hole to hard ones. The hole should be foreseen with a countersinking equalling the diameter of the screw in diameter and depth.

Threaded Holes

Threaded holes, a recurring feature of industrial components, can be formed during moulding if start of the thread is placed well below the surface to avoid damage. Likewise, the thread should stop a few turns before the end (Fig. 109).

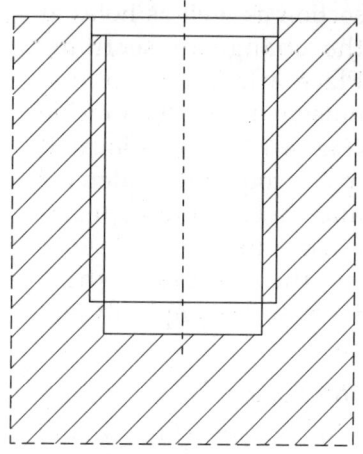

Fig. 4.109

Assembly Aids

Assembly aids; snap fit joints for temporary or permanent attachment of moulded components facilitate assembly without additional fasteners. Their design depends upon certain properties of the plastics employed. Figure 4.110 depicts an

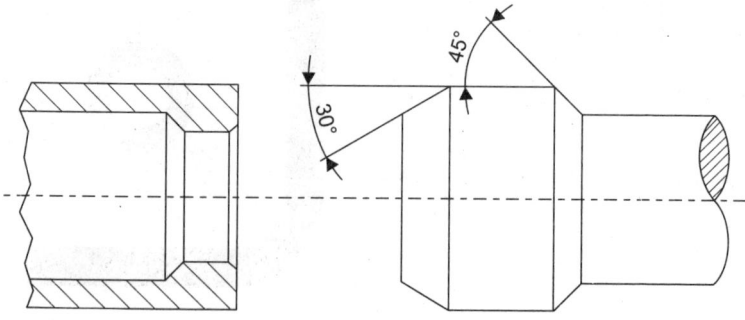

Fig. 4.110

example of a permanent snap-fit joint. The hole, which is smaller than the shaft, is shaped to guide the entry of the later. The hole expands within the elastic limits to permit entry of the bigger part of the shaft and resuming its original size thereafter, holds it permanently in its narrower portion or neck. In assembled state, neither of the two parts is under stress.

Integral Hinge

Integral hinge in polypropylene makes it possible to produce a container and its lid as one moulding. A properly designed and properly moulded hinge has virtually an unlimited life. Figure 4.111 illustrates the main features and dimensions. Polypropylene acquires a high orientation by being flexed which increases its tensile yield strength enormously.

The pp-hinge is basically a light duty feature, limited to applications where bending force is the main load.

Integral hinge can also be moulded in polyacetal.

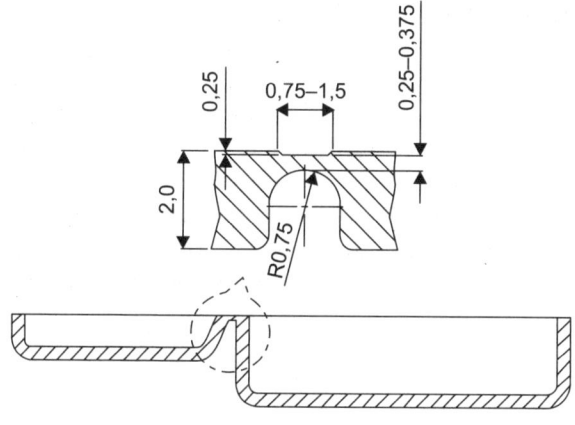

Fig. 4.111

Apertures and Side Holes

These features require moving components in a mould, limit the number of cavities which can be accommodated and add to the cost. A modification in their design as shown in Fig. 4.112 can make it possible to form them without the moving aids in the mould.

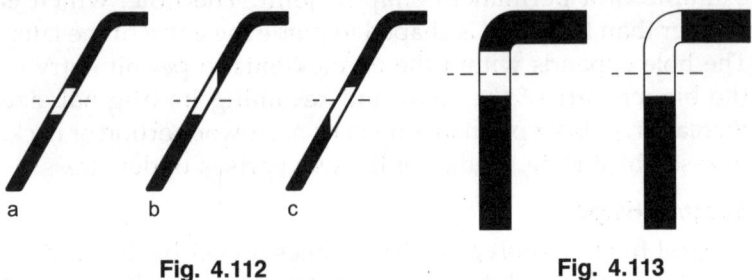

a   b   c

Fig. 4.112   **Fig. 4.113**

Figure 4.113 shows yet another example of redesigned side opening.

Additional holes, apertures and cut outs, if located judiciously, can help reduce the weight of the article. It must, however, be examined whether they give rise to unacceptable weld lines or weak spots. In certain type of products, the cut outs may even add to the aesthetic appeal of he product.

Insert Moulding (Fig. 4.114)

Metallic inserts, such as reinforcing metal plates, electrical contact pins, internally threaded metal sleeves, etc., should not be too large as compared to the moulding. The thin plastics material around a big insert, when prevented from shrinking, may develop cracks. They should not be positioned near the surface or the edges. Provide flats, knurling, and grooves on the outer surface of round inserts to prevent turning and withdrawal. These undercuts should not have sharp edges.

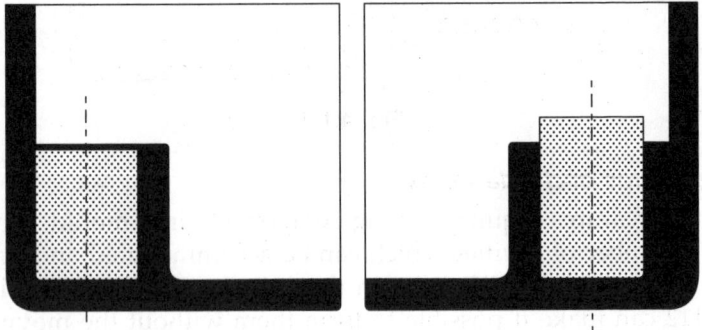

Fig. 4.114

Tolerances

It is very difficult to predict shrinkage precisely, especially for materials with a wide shrinkage range and for the articles with varying wall thickness. Very narrow tolerances add to the cost of the mould and cause more rejection in production because any variation in material, ambient conditions and moulding parameters influences the shrinkage.

Tolerances can be narrower for thermoplastics with less shrinkage. They should be larger on the free dimensions of the product. More tolerance should be allowed if the mould has to be a multi-cavity one.

Containers for household as well as for industrial use were the first articles where thermoplastics proved their worth. Soon after, housings for household gadgets, radios, gramophones etc. were moulded in thermoplastics. It would be noted that many of the above guidelines relate to such products and are the outcome of an experience extending over a century. Now, plastics are finding application in all major technical fields. The new applications have given rise to additional design rules. However, the principle that a good design involves the use of minimum of material so distributed as to ensure sufficient strength at all points, remains unchanged.

Before finalising design of a product in plastics, it should also be reviewed from the tooling angle. Is there a feature which makes the tool very complicated? Will the mould design necessitate very delicate mould components prone to frequent damage, repair and replacement? Can the design be modified to make the mould simpler without influencing its function adversely?

Mould Making

MOULD MANUFACTURING

Although there has been a lot of progress in the field of mould making, certain basic realities have not changed. A mould may be said to consist of two main parts, viz. the bolster and the core and cavity. Conventional machining methods have retained their position as the most economical way of making the bolster and even the core and cavity for regular geometrical shapes. Turning, drilling, milling, surface and cylindrical grinding, jig grinding and polishing operations are even today as irreplaceable as in the early days of mould making. The corresponding machine tools have become more sophisticated, efficient and accurate. CNC technique has brought speed, economy and reliability but the basic concept of material removal has remained unchanged. The conventional machine tools still form an important and sizeable part of a mould making tool room and they still do the bulk of material removal.

It cannot be overlooked that the process of EDM has revolutionised the basic concepts of core and cavity fabrication. Almost all constraints to the shape and form of the article to be moulded have been removed with the advent of spark erosion and its allied process, viz. the wire cutting. Since their introduction, plastics have been able to enter all fields of industry.

226

It is not intended to dwell upon the conventional processes well established in this field. In fact, these metal removal processes and machine tools are not peculiar to the field of mould making only. It is intended to confine the dissertation to certain specialised techniques, which are not found in every tool room but are still useful in out-of-the-way cases.

Special Mould Making Processes

Ceramic Casting (Fig. 5.1)

The manufacture of core and cavity in steel by the conventional process of sand casting is unsuitable due to poor surface finish, porosity and inaccuracy. However, the ceramic casting process is not plagued by these drawbacks and is often employed for making cores and cavities of large moulds.

The first step is the fabrication of a pattern of the mould core with final dimensions including the shrinkage allowance for plastics as well as steel. The pattern is positioned in a casting box and slurry consisting of fine refractory particles suspended in a bonding agent like ethyl silicate is poured over it (Fig. 5.1A) The slurry block, after solidification, is removed from the casting frame, ignited and then placed in a furnace at 900°C to remove water and alcohol (Fig. 5.1B). The negative ceramic mould is now ready to receive molten steel at 1600°C (Fig. 5.1C). After solidification, the mould component (core) is ready for further finishing operations. The achievable dimensional

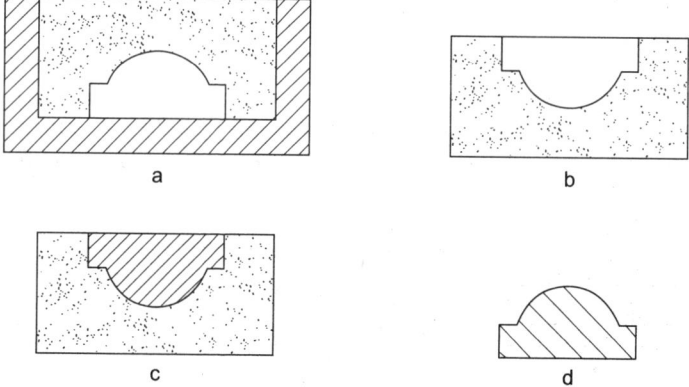

a

b

c

d

Figs 5.1a to d

accuracy is about 0.5%, which may be found adequate in many non-critical cases.

The cavity block can also be manufactured in the same way. The pattern can be made of any material which yields a smooth, non-porous surface and does not react with the ceramic slurry.

Epoxy Resin Casting (Fig. 5.2)

Moulds meant for a limited production can be readily cast out of special epoxy resins, impregnated with up to 80% aluminium powder, which lends them strength, polishability and heat conductivity. The model for casting can be made out of metal, wood, plaster, leather, silicone, plastics resin, etc. The model is positioned in a casting frame; metal inserts, if any, can be fixed onto it and even the cooling pipes can be placed before the preheated and degassed mixture of resin and hardener is poured over the model. Both halves of the mould are cast in this fashion separately, left to cure for a day, taken out of the casting frames, assembled and heat cured. The mould inserts, cast in this fashion, can be fitted in a steel frame. The ejector holes and the runner system can be machined in the conventional way.

The resin reproduces all details of the model faithfully. It can also be highly polished. The resin has negligible shrinkage (0,02%), having practically no effect on the product dimensions. The cast resins have adequate surface hardness but low tensile strength. Such moulds may yield over 1000 shots, depending upon the intricacy of the product and the plastics material being moulded.

Electroforming

It is an electro-chemical process of material deposition on a given pattern, which has the reverse form of the required

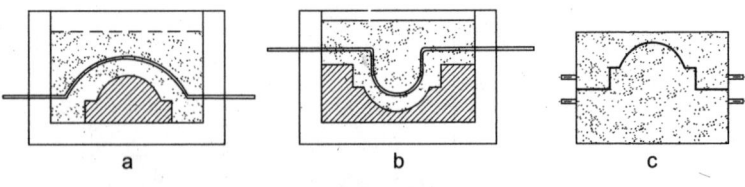

Fig. 5.2

cavity/core. Because of the ease of fabrication of a pattern or model in male form, the method is generally adopted for production of cavity inserts.

The model can be made out of any easily machineable material such as brass, acrylics or epoxy resin. However, the non-conducting materials have to be provided with a conducting outer layer, mostly a very thin coating of silver.

The pattern acts as a cathode in an electrolytic bath and nickel-cobalt is deposited on it till a shell of about 4 mm. thickness is formed. The Ni-Co shell is backed up by electrically deposited hard copper till the block achieves a size out of which an insert can be machined (Fig. 5.3).

Although it is a relatively slow process, it has the advantage of yielding inserts with adequate hardness (~45–55 Rc.). The surface finish too is as good as that of the pattern. It reproduces the accuracy of the pattern, which though not hardened, can be used again and again.

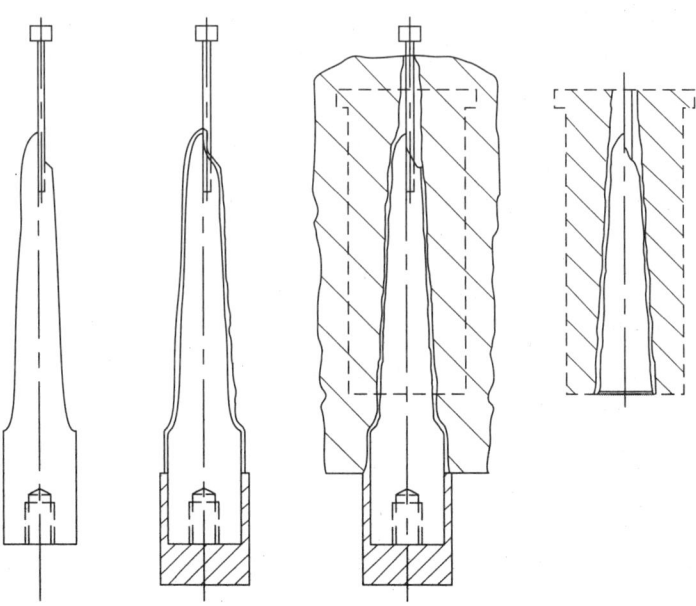

Fig. 5.3

Electroformed inserts are used for deep, narrow cavities such as those for textile garn cones, medical disposables, etc. which would be difficult to polish if made by other processes.

Cold Hobbing (Fig. 5.4)

Cold hobbing is a technique of producing cavities by metal displacement instead of metal removal. A hardened and polished punch called hob, which has the external contour of the intended cavity, is slowly forced into a block of soft annealed low carbon steel, positioned in a conical holder, at room temperature. The depression achieved has the shape and surface quality of the hob.

Whereas the hob has the exact dimensions of the required cavity, the steel block is considerably bigger than the final size of the cavity insert. Its diameter and thickness is not less than twice the diameter and the depth of the impression to be sunk. After hobbing, it is machined down to the final size of the insert and heat treated.

The process is particularly economical for multi-cavity moulds with cavity forms which may be difficult/ expensive to machine and polish but can be conveniently produced as a

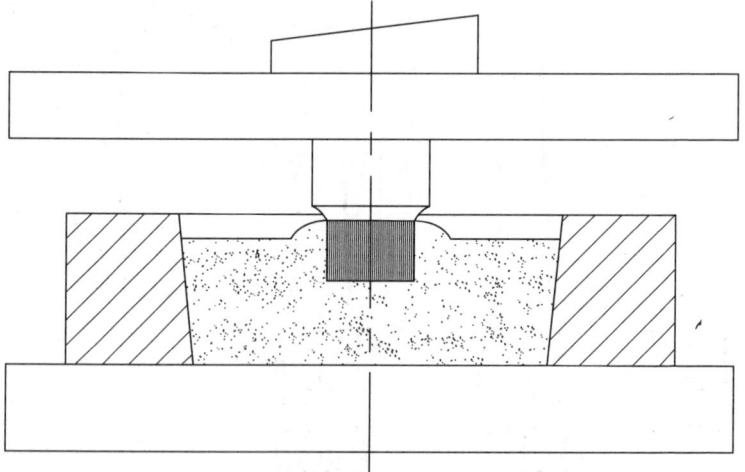

Fig. 5.4

punch. One hob can produce any number of cavities, identical in shape, finish and dimensions. The finish of the hob is reproduced faithfully in the sunken impression and little polish is needed after case hardening of the insert. The structure of the material also becomes more compact due to compression. Another advantage is the creation of raised lettering on the horizontal or nearly horizontal surfaces of the hobbed cavity.

Metal Spraying (Fig. 5.5)

A very fast method of producing a cavity/core is the spraying of low melting temperature alloys of bismuth and tin with the help of compressed air onto a pattern possessing the reverse shape of the cavity/core. The pattern can be made of metal, ceramics, wood, epoxy resin or even thermoplastics. In other words, even a sample can be used as pattern.

The pattern is treated with a releasing agent and the alloy is sprayed on it till a shell of about 4–5 mm. thickness is formed. Though the metal may have a temperature of 200–400°C, it cools down to 40–50 degrees when it hits the pattern and settles down on it in the form of a non-porous layer. The shell is backed up by epoxy resin impregnated with aluminium powder to form an insert, which can be fitted in a steel bolster. It is also possible to incorporate cooling by placing bent copper pipes in position before pouring the backing material.

The bismuth alloys are comparatively soft. The life of an injection mould is limited to about 1000 shots. However, the

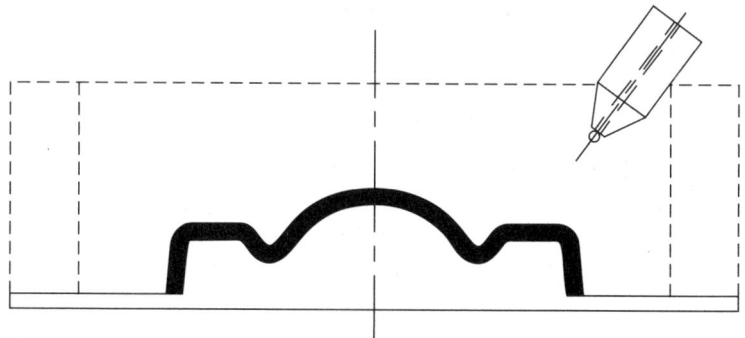

Fig. 5.5

speed of mould making enables quick production of prototypes or small pilot series.

The expensive bismuth alloys can be salvaged and reused.

Chemoforming

It is a process of selective metal removal with the help of acids. Its working is similar to that of EDM or spark erosion without the use of electric current for removing the metal. As in the EDM process, here too a movable horizontal table with a tank is required to house and position the job or the insert. The surrounding fluid is an acid. The pre-machined job is coated with a protective material. The "electrode", held in the upper plunger of the machine, is the negative replica of the desired shape. It can be made of plastics or any other material as it is not subjected to any wear or tear. Its function is to touch the work-piece and remove the protective coating. The work of metal removal is accomplished by the acid. Through repeated strokes, more and more surface is bared, more and more metal removed and gradually the job gets a contour which is an exact reverse reproduction of the "electrode" in dimensions and design details and surface finish. If the "electrode" is well polished, the job also does not require any further touch up.

The process is faster than EDM. Its significant advantage is that it does not effect the structure of the metal. Also the "electrode' does not undergo any wear and tear and can be used any number of times. Furthermore, the workpiece may be hardened or unhardened and may be made of non-ferrous metals like copper, brass, bronze and aluminium. The surface finish is finer than that achievable by spark erosion.

The process proves highly economical in conjunction with other material removal processes like machining and EDM.

Direct Metal Laser Sintering

This is the latest process of forming an article or a mould insert with the help of a laser beam, which is directed and moved by 3-D data on fine metal powder. A thin layer of the metal powder melts under the heat of the moving laser beam and sticks together in the shape generated by the beam. Layer by layer, the desired article or a mould insert is created out of the metal powder. Its physical properties such as tensile strength, hardness, etc. depend upon the composition of the metal powder.

Out of a variety of metal powders available for the process, those most suitable for injection mould inserts are either a bronze-based mixture or a tool steel one. The former mixture is faster in manufacturing but is comparatively softer (115 HV) and is generally employed for moulds for trial series. It is, however, possible to mould up to a few hundred thousand components depending upon the thermoplastics to be processed. The maximum operating temperature is 400°C.

Inserts, laser-sintered out of the bronze mixture, display a minimum rest porosity of 8%. The surface is generally sealed by micro shotpeening before use. The redeeming feature is its good heat conductivity, which results in faster heat dissipation and shorter moulding cycles.

A tool steel material comprising Marging 300 (DIN 1.2709 - X3NiCoMoTi) is suitable for inserts for production moulds. In as-sintered state, the inserts possess a hardness of 35–37 Rc but can be easily hardened up to 50–54 Rc. In hardened state, their service life is comparable with those manufactured in traditional fashion. They can withstand temperatures as high as 1100°C. They are almost fully dense (porosity lower than 0.5%) and do not require a surface sealing treatment. The surface lends to a good finish through polishing.

Laser sintered inserts are generally incorporated in mould bolsters but up to a certain size, they can replace mould plates and form a mould directly. Figure 5.6 shows a laser sintered mould insert along with the moulding.

Fig. 5.6

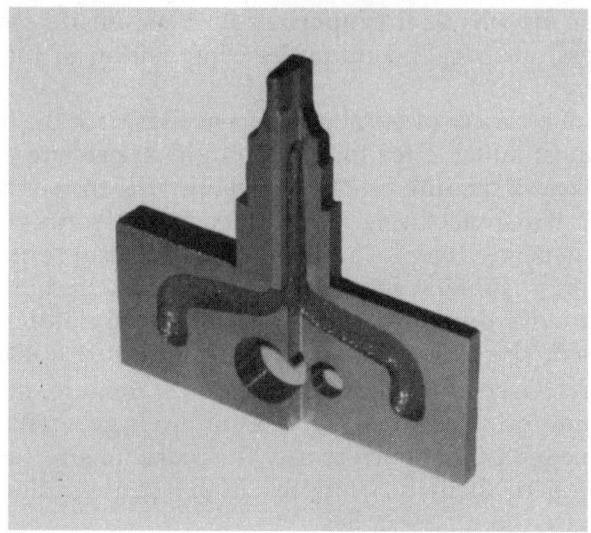

Fig. 5.7

A unique advantage of the process is the speed of manufacture without any model or pattern and the facility of incorporating integral cooling network, matched to the contours of the product as well as ejector holes and even undercuts during laser-sintering. The integrated 3-D cooling network leads to faster production, more uniform shrinkage, less stresses and consequently lower rejection rate (Fig. 5.7).

The Direct Metal Laser Sintering is in the process of further development to enable manufacturing of moulds out of different alloys with longer production life.

MATERIALS FOR INJECTION MOULDS

Injection moulds are subjected to severe conditions such as fast fluctuating temperatures, very high pressures with cyclic changes, wear, corroding fumes and rust. They are required to maintain high dimensional accuracy and reliable functionability under extreme working conditions. This is why, the production moulds are fabricated out of alloyed steels, alloying elements like Chromium, Vanadium, Nickel, etc. adding the desired property. Depending upon the function of the component, the

required finish and the working life, following steels has proved their suitability for moulds.

In order to select the right material for fabrication of mould components, one must consider conditions of operation, methods of their fabrication, demands on quality and accuracy, the working life and economics.

A mould is, as a rule, a one-time job, a prototype so to say. The most economical method of its fabrication is machining. In other words, the cost of fabrication is directly linked with the time required for machining. Easy machinability will, therefore, be a positive feature of the material.

The mould bolster is subjected to compression from the clamping force of the machine and to deflection by the injection pressure. The guiding elements and the moving components undergo friction and are likely to wear out, causing inaccuracy and moulding defects unless made of a hardened material. Depending upon the climatic conditions and the plastics being processed, the mould parts may be attacked by rust and corrosion. Cooling water can give rise to rusting from inside the mould inserts, which may result in creation and propagation of cracks.

The components shaping the moulding are required to serve for a long time under stringent conditions of quickly fluctuating pressure and temperature. Plastics melt, flowing at a very high speed, subjects the material to wear and tear and dimensional variations. In case of filled plastics, this effect is very prominent. The surface finish of the product is a direct replica of the mould surface and in most cases high gloss is demanded. Last not least is the property of heat conductivity which directly influences the cooling time and consequently the productivity.

In the light of above considerations, it may be summarised that the mould materials should possess the following attributes:

a. Easy machinability
b. Toughness
c. Fatigue resistance
d. Resistance to wear
e. Strength at elevated temperatures

f. Hardenability without distortion
g. Good polishability
h. Corrosion resistance
i. Good heat conductivity

Ferrous Materials

Unalloyed Steels, e.g. EN-8

It is the most common steel employed for mould bolsters and many other components not coming in contact with the melt and not subjected to friction. Generally, it is used in unhardened state and has a tensile strength of about 60 kp/sq. mm. It is easily machineable but is not rust proof. In exceptional cases, it may also be hardened.

EN-8 is not recommended for the mould components forming the product. However, for cheap moulds with a short life, it may be used with a proper surface treatment like chemical nickel plating, hard chrome plating, nitriding, etc.

Prehardened Steels Like 1.2311, 1.2312 (P20, P20+S)

The prehardened steels are used for making bolster for high production moulds and also for core and cavity inserts of moulds with medium range of life and accuracy. Their main advantage lies in their initial toughness (tensile strength 110 to 140 kp/sq. mm), obviating the need of hardening operations, which may lead to distortion. In rare cases, the surface hardness is enhanced by nickel/chrome plating and nitriding, etc.

Due to initial toughness, the prehardened steels are more difficult to machine. The machinability can, however, be improved by incorporation of sulphur in the structure. It has, however, adverse effect on the surface finish, especially if any surface structuring is to be carried out.

Their polishability is adequate for industrial and less demanding articles.

Case Hardening Steels DIN 1.2162 (AISI P2), DIN 1.2764 (AISI P21)

These steels are used for mould bolsters of high production tools and also for mould inserts such as cores and cavity inserts for relatively less complicated large products. The machining is easy but hardening operation is unavoidable. Therefore, some after-operations like grinding and repolishing must be carried out. These materials are economical in price, have good polishability and long service life. Through nickel or chromium plating, the resistance to corrosion can be improved. Case hardening steels, when properly annealed, can also be employed for cold hobbing—an economical process to produce identical mould inserts for multi-cavity moulds.

Because of their hard surface and tough core, case hardening steels are also used for fabrication of mould components subject to wear and bending.

Through Hardening Steels DIN 1.2767 (AISI GF7)

The advantage of through hardening steels lies in the uniformity in their structures through and through. Thus they have less in-built stresses, negligible distortion and long service life. The steel cited here has excellent dimensional stability and polishability. It is the most widely employed material for inserts of the high production moulds. This steel also lacks the corrosion resistance and must be surface treated in critical cases.

Hot Die Steels DIN 1.2343/H 11

Hot die steels possess a tough core and retain their strength and hardness also at elevated temperatures. They can also withstand temperature fluctuations. They are highly suitable for making hot runner manifolds and components of moulds for plastics such as speciality polymers needing high mould temperatures. They can be machined, spark eroded, hardened and polished. Nitriding is the preferred hardening process for hot die steels as it gives them a hard, wear resistant surface and does not reduce the core toughness. They are not corrosion resistant but can be plated.

Nitriding Steels (EN-31)

Nitriding is a process of surface hardening at comparatively lower temperatures, by incorporating nitrogen into the steel. It forms hard nitrides with some elements like Al, Mo, Va, and Cr. The main advantage lies in the absence of distortion due to hardening. No after-operations are, therefore, required for correction. Usually the steels are prehardened before nitriding to combine a tough core with a hard surface. Nitrided surface also forms a rust resistant layer. Sometimes all plates of the bolster of an expensive mould are nitrided solely as protection against rusting in tropical climate.

Stainless Steels DIN 1.2083 (AISI 420)

Some plastics release aggressive gases like chlorine, formaldehyde etc. on overheating, which attack the metals and corrode them. PVC and POM (Polyacetal) are the prominent examples of such thermoplastics. Chromium plating of moulds/inserts solves the problems to some extent, but the chromium layer tends to peel off after a little damage. Secondly, the thickness of the deposited layer is never uniform which may be found unacceptable in case of very tight tolerances.

Corrosion is also caused by some additives in certain plastics. High moisture in the air may condense on the cold mould surface leading to corrosion. The cooling water is another potential source of rust. The sharp furrows in the drilled cooling channels develop into internal cracks under the effect of rust.

Stainless steels for moulds inserts offer a good solution to these problems. The chromium content in the alloy steel has to be 12% or above. It forms a protective layer of chromium oxide on the steel, which prevents corrosion once for all. These steels are hardenable to about 52–56 Rc.

It may be added that the steels containing chromium are inferior in conduction of heat.

Non-ferrous materials

Copper-beryllium Alloys

Cu-Be alloys, especially those with a beryllium content of more than 1.7%, combine excellent heat conductivity with adequate hardness (35-38 Rc) and toughness. They represent an ideal material for the mould components like slender cores and thin inserts, which cannot be cooled directly. Be-Cu can be finely polished and is corrosion resistant which property can be further enhanced through chemical nickel plating. Their gliding

characteristics are also better than those of tool steels so that trouble free ejection can be achieved with less taper.

Cu-Be is also used for manufacturing heat conducting nozzles of hot runner moulds.

Aluminium Alloys

Certain aluminium alloys can be used for short-run moulds. The Al Zn Mg Cu 1.5 (DIN 3.4363) alloy, which is employed in aircrafts, has very good heat conductivity and a modulus of elasticity equalling a third of that of steel. Apart from a good cooling effect, these alloys have the advantage of easy and fast machinability—also through spark erosion and adequate polishability. Due to their lightness, the mould parts can be transported in the tool room without the help of cranes. However, their use remains confined to trial injection moulds or those for pilot series.

Tin-bismuth Alloys

These are low melting temperature alloys (melting point about 200°C) which can be sprayed. Mould inserts can be formed within a matter of hours, by spraying the molten metal on a model. The reproduction of details is very good. The inserts can be mounted in standard bolsters made of steel to make experimental moulds within 24 hours. It must, however, be underlined that the mechanical strength of bismuth alloys is very low. Hence the moulds primarily serve as a fast help to get prototypes for trials.

Bismuth alloys are quite expensive but they can be remelted and reused any number of times.

Bismuth alloys are also employed to make cores for the "fusible core" moulding technology (Chapter 10).

POST-TREATMENTS

Heat Treatment

As a rule, the product-forming components of a production mould, viz. the core and cavity, are hardened to prolong their service life. It is also not seldom that the plates housing these mould components are also hardened and ground out of the same consideration. Guiding members, and other moving components like ejectors, slides, strippers, pullers, etc. which are subjected to wear, are invariably hardened. All conventional heat treatment processes such as case hardening, through hardening, nitriding, vacuum hardening, etc. are also employed for injection moulds and their components. In fact, most standard components come in hardened state, ready for use.

Polishing

The mould components, shaping the moulding, leave their imprint on the surface of the product. Thus the marks left by the machining operations on core and cavity of the mould are transferred onto the product. In most cases, these are not acceptable. The uneven surface may also come in the way of easy demoulding. This is why, the relevant surfaces of all such components are worked upon or "polished" to eliminate the machining marks before hardening. The degree of polish depends upon the end use of the moulded product. Polishing of moulds for technical components is usually confined to removing all traces of machining. However, for the products in which the surface finish has an aesthetic value, the polishing may call for a smoothness of 0.001–0.01 µm. The mould cores and cavities for consumer products are required to have a mirror like surface finish. The relevant surface are polished again after hardening to remove deposits left by the hardening process.

The process of polishing is essentially manual and is carried out, with or without the help of electrical hand tools, in a number of steps. The first step consists of removing machining marks with corundum polishing stones of different fineness in sequence and then going over to lapping with emery pastes. Here too, one starts with relatively coarser grit size, coming to very fine sizes in two or three steps. Mostly pastes with fine diamond particles are used in final stages. For technical plastics, pastes used are P 240–320. However, for shining surfaces of opaque articles, the range will be P 240–400 and for transparent mouldings, final polish with P 500 is imperative. In most cases, polishing is done or "drawn" in the direction of ejection for easy demoulding.

A precondition for obtaining a mirror like polish is corresponding steel quality. It must be free from inclusions and have fine grain structure.

The process of electro-polishing is based on removing the top layers micronwise from the surface to be polished electrochemically. The surface must, however, be absolutely free of machine marks and unevenness which would otherwise get accentuated. The process can deliver mirror like finish on prepared surfaces. The steel too must be of a suitable quality.

Most common mechanical tool for polishing is the flexible shaft grinder. Polishing can be carried out with the help of special machines in some cases where the surface is large and regular. The operation is purely manual for very small cavities with sharp corners. Fine polishing without the help of mechanical devices may take about half to one hour per square centimetre.

The ultimate quality of the polish is also dependent upon the homogeneity of the steel structure, the extent of impurities, the grain size and the heat treatment.

Some products are designed with matt surface finish in parts or completely. Here too, the mould components are first polished to remove the machining marks and then finished to the required degree of polish. Ultimately, the designated portions are roughened through sand or glass bead blasting, spark erosion etc, after masking the rest area. The process of spark erosion is capable of delivering surface of desired roughness. Silken surface produced by spark erosion is found to be the most suitable for polymers containing glass fibres as it camouflages surface defects like dull spots.

Plating

The mould surface can be made more durable and more resistant to aggressive materials through plating with chromium or nickel. Hard chromium plating is a galvanic process, carried out in a bath containing chromium solution. The component to be plated acts as a cathode. The chromium layer lends additional hardness to the mould surface. Moulds for polymers like PVC or polyacetal, which may release aggressive chemicals upon overheating, are provided with a protective layer of chromium. However, it does not undergo a bond with the parent metal of the mould and may peel off if damaged. The thickness of layer may not be uniform and deep, blind holes may remain partially unplated. The article may have to be ground after chromium plating for greater accuracy.

Nickel plating, on the other hand, can also be carried out electrolessly in a chemical solution containing nickel and phosphorus. The object to be plated is cleaned thoroughly with acids and alkalies to remove rust, scaling and oil and rinsed frequently in between. A layer of nickel is deposited chemically without the use of electric current.

Electroless Nickelplating offers the Following Advantages

- It forms a bond with the surface and cannot be peeled off.
- The thickness of the nickel layer is uniform all over, also in the holes.
- The plated layer is hard, wear, galling and corrosion resistant.
- It can yield matt, semi-bright or bright finish.
- It needs no electic current.
- The equipment is inexpensive.

PTFE-Coating

The process is similar to the electroless nickel plating and is carried out in a chemical solution containing nickel, phosphorus and PTFE in the form of microscopic beads. The deposited layer, 12–200 micron thick, may contain 20–25% of the polymer.

PTFE in conjuction with nickel provides a hard surface with a coefficient of friction as low as 0.1. In addition, it possesses all other characteristics of electoless nickel plating. The coated layer is hard, wear and corrosion resistant and has good adhesion with the parent surface. It provides dry lubrication with low friction which is essential for safe ejection of mouldings with insufficient draft.

The nickel-PTFE coating can be carried out on cast iron, steel, tool steels, stainless steel, Be-Cu, aluminium, etc. The PTFE coating of crucial mould parts enables easy release with less taper, reduces the moulding cycle and extends the mould life.

It can be used in moulds for thermoplastics, thermosets, liquid silicone rubbers and polyurethanes. The surface hardness of about 30–45 Rc is found to be adequate in most cases.

Structuring

Quite often, the consumer products are provided with granular surface to look like leather, wood, fabric, etc. The surface of corresponding mould components is formed accordingly. The process resorted to is called photo etching. The polished mould surface is coated with light-sensitive chemicals and an image of the intended surface design is exposed on it. It is processed like an exposed negative and then treated with etching chemicals which act on the surface according to the intensity of light and shade of the exposure. A negative structure with varying depth is created by the corrosive chemical.

Photo etching is a corrosive material removal process. This is why, stainless steels with high percentage of chromium are difficult to etch.

Tool steels can be hardened before photo etching but the thickness of the hardened layer must exceed the depth of etching. Again, steels with sulphur are not recommended for photo etching treatment as sulphur may get washed out and leave pits on the surface.

Non-ferrous metals like aluminium, copper and zinc can also be etched and chromium plated.

The structured vertical surfaces act as deterrent to ejection. Such moulds need larger draft angles for safe demoulding. As a rule, a taper of 1° must be provided per 0.1 mm. depth of etching.

Intrusion Injection Moulding

THE PROCESS

Intrusion injection moulding may rightly be regarded as an extension of the conventional injection moulding process as it can be carried out with the same machine without hardware alterations and without additional equipment. The process is a combination of extrusion and injection moulding as the filling of the mould is achieved partly by extrusion and partly by injection. In other words, it is a combination of low pressure and high pressure injection moulding steps. The process is carried out in four steps:

- The mould is closed and the nozzle of the plasticising unit is brought in touch with the sprue bush of the mould. The screw is in the forward position, close to the nozzle, i.e. the barrel is empty.
- The charging or dosing operation is started with back pressure. The melt fills the mould gradually, the back pressure acting as the injection pressure. Here the rotating screw is basically acting as a conveyor of the melt as in an extruder, filling the mould with low pressure. As the mould gets filled, the back pressure forces the rotating screw to recede to a pre-set point and stop dosing.
- The screw takes over the function of a plunger, makes a forward stroke and compacts the filled material. A holding phase follows.
- After cooling, the moulding is ejected out in the normal manner.

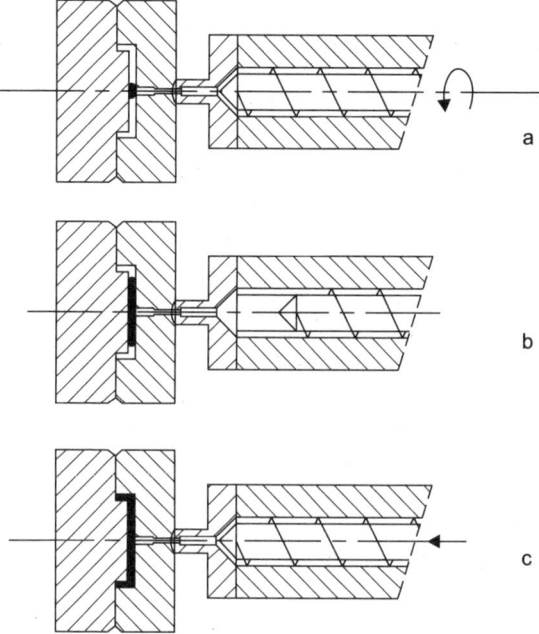

Fig. 6.1: Illustrates the process in three steps

An alternative to the above procedure is as follows:
- Charging during the cooling time till the screw has arrived at the pre-set end position.
- Filling the mould through extrusion, i.e, through rotation of the screw in the end position after the mould has closed again after opening and ejection.
- Ultimate injection through forward stroke of the screw as plunger.
- Holding and cooling as in conventional injection moulding process.

Apart from the two variations described above, a number of alternative combinations are possible.

THE EQUIPMENT

The moulding machine for intrusion moulding process differs from the conventional one only in a few controls which permit extrusion for the set time and subsequent injection and holding.

Modification in controls to effect pulsating holding pressure help eliminate surface defects.

A normal injection moulding machine can be manipulated to perform the process in manual mode without any alteration. The method is resorted to only for trials or for production of a very small lot.

THE MOULD

The mould for the intrusion moulding process is identical to the one for the conventional injection moulding. The difference lies in the dimensions of the feed paths which, understandably, have to be larger than those for normal, high pressure moulding to permit passage of melt at lower pressure. It is also advantageous to maintain the maximum permissible mould temperature for the polymer in question.

ADVANTAGES AND DISADVANTAGES

The advantage of the process lies in overstepping the limitations of maximum shot volume. Depending upon the wall thickness of the article, products with ten times the nominal shot weight of the machine can be injection moulded by intrusion injection. The safe wall thickness to length of flow ratio is 1:50.

The low pressure may give rise to sink marks or blow holes in case of thick walled mouldings. Machines with arrangement for pulsating holding pressure help overcome the problem.

It is possible to mould elastomers as well as thermosets by this process.

The process has a limited application because it can essentially mould thick walled articles. The machine controls, too, must be altered. The production is slow.

APPLICATIONS

Trial and operation of moulds of thick walled articles on smaller machines.

Moulding of thick walled articles like boxes, flower pots, furniture components.

7

Injection Compression Moulding

- The process
- The mould
- Application
- The Equipment
- Advantages

THE PROCESS

As the name suggests, the process is a combination of injection and compression moulding. The aim is two-folds. It exploits the efficiency of the injection moulding process but employs compression to even out the differences in dimensions brought about by differential shrinkage unavoidable in injection moulding. A typical example of products of the process are optical lenses moulded out of thermoplastics.

Another advantage of the combined process is the possibility of moulding large articles with high L:T (length of travel of the melt/wall thickness) ratios. Under normal circumstances, it may not be possible to fill the cavity on the particular machine. The very high pressure needed to fill the cavity may force open the mould, but if the injection takes place with partially open mould which amounts to an increased wall thickness, injection of the amount of melt required for the final product can be accomplished with less pressure. The machine can close now and compress the melt to its final thickness, pushing it simultaneously to the unfilled outer areas. The cooling takes place under a pressure acting all over the surface.

Basically, it is an injection moulding process followed or accompanied by compression. There are four variations of the process depending upon when the second step, that is the compression, is brought into action.

Sequential Injection Compression Moulding

As the name suggests, the two operations of injecting and compressing take place one after the other. The mould is closed incompletely, leaving a gap of twice the final wall thickness; a measured amount of material is injected and then the mould is closed to its final position, thereby compressing the melt all over uniformly (Fig. 7.1).

Because of the increased wall thickness, injection takes place with less pressure. Consequently, the locking force needed to counter the pressure is also low. The moulding can take place on a smaller machine as compared to the standard injection moulding. As the mould closes and the compression sets in, the melt is pushed to the unfilled extremities. In between, there is a phase of "hesitation" when the melt was stationary. The restart of melt movement bestows flow lines on the surface of the product, which may not be acceptable in some cases.

The injection pressure remains below 75% of that needed for normal filling and the clamping force does not exceed 30%.

Simultaneous Injection Compression Moulding

In this variation too, injection starts with the partially open mould but the closing of the mould commences almost simultaneously, with a maximum lag of 2–3 seconds. The result

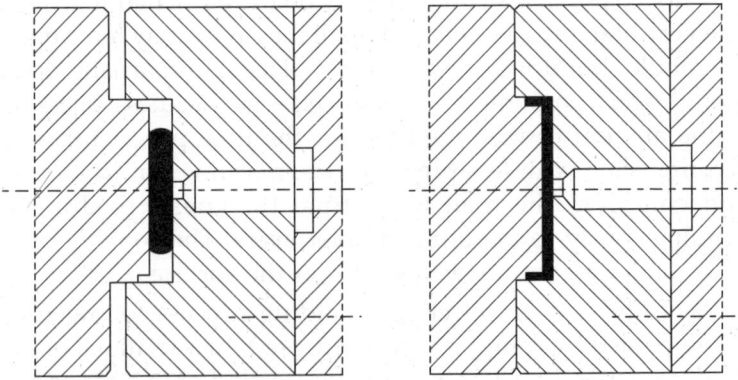

Fig. 7.1

is that the melt front remains constantly in motion and has no "hesitation" pause. Consequently, there are no surface marks.

Both variations suffer from a common drawback. In case of large mouldings, the melt tends to flow downwards under the force of gravity, causing the lower side of the mould to fill first. This may introduce problems of uneven distribution during compression.

The injection pressure and the locking forces are low like in the foregoing case.

Breathing Injection Compression Moulding

The injection starts with the mould closed to the final position but it keeps on opening continuously, either under the injection pressure or through pre-programmed controls. The melt remains always under pressure. This eliminates the possibility of uneven filling encountered in the previous two variations of the ICM process. At the juncture, when nearly complete volume has been injected, the clamping force is activated and the melt is compressed to the final wall thickness.

Passive Injection Compression Moulding

It may be rightly regarded as a variation of the breathing injection compression moulding applicable in special cases and practicable with machines having fully hydraulic locking.

The process makes use of controlled breathing of the mould during injection by setting the locking force lower than the internal mould pressure arrived at after complete filling. It is effected by first filling the mould under normal locking force of the machine and then reducing it gradually till the moving half recedes by an amount permitted for venting in case of the particular plastics. As the mould opening force generated by injection decreases because of partial cooling of the injected melt, the machine closes fully under its set locking force.

The process does not require any modifications in the machine hardware. The mould is equipped with sensitive dial gauges to display the gap between its two halves for accurate adjustment of the locking force. The fully hydraulic machines are most suitable for the process as hydraulic oil is compressible.

An essential condition of the process is that the injected volume of the melt must remain constant irrespective of the mould breathing. It can be achieved by adjusting the charging to the amount needed for the article. The screw must inject the full amount of the dosed melt and touch the end of the cylinder. It must be held in this position with sufficient force so that the melt being compressed through the closing action of the moving platen cannot force its way back into the injection unit.

The process is suitable for large round products with central gating as the side gating may cause partial flashing during breathing.

Selective Injection Compression Moulding

By selective ICM, the mould is completely closed before injection and a separate core in the mould is moved up locally, during or after injection (Fig. 7.2). This results in localised compression, compaction and thinning in a relatively large section. In some cases, the compression pressure exerted by the core can replace the holding pressure albeit with the difference that the compensation for shrinkage takes place not by pushing more melt in the cavity but by displacing a part of the available material. If necessary, the holding pressure can be active along with the compression process.

The selective ICM can be carried out on a conventional moulding machine with a core pulling device. In absence of

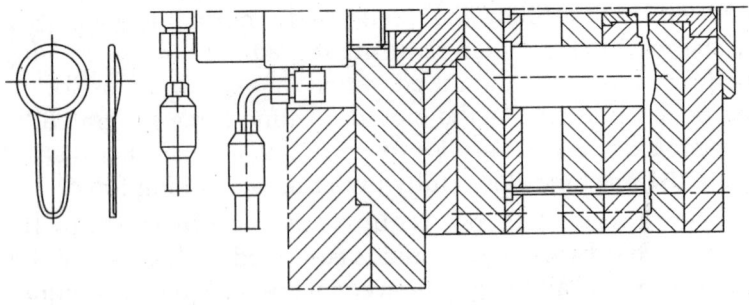

Fig. 7.2

this option, a separate hydraulic facility can activate the core. The connections to the hydraulic ejector, with alteration in machine controls, can also be employed to operate a hydraulic device for movement of the core/cores after injection in the still closed mould as shown in Fig. 7.2.

Injection compression moulds for lenses work according to this method with the difference that not a part of the core, but the whole core forming one side of the moulding, is pushed up with the mechanism built in the mould and operated with core pulling device of the moulding machine.

Selective ICM can also be gainfully employed to eliminate sink marks in areas away from the gate, where the holding force may cease to be effective before solidification due to freezing of the gate.

THE EQUIPMENT

The Moulding machine: The injection moulding machine performing injection compression process differs from a conventional moulding machine only in a few controls governing the mould closing and the injection operations. The mould closing unit of the machine should stop at a predestined point and have the set closing force to counterbalance the injection pressure. In the second step, it should close fully after the preset duration and develop full locking force. It should be possible to regulate the speed of closing. This is called the sequential compression stroke. As an alternative, the second movement can also take place at the time of injection. This variation is termed as the simultaneous compression control. Likewise, it should be possible to have the moving platen open with controlled speed during injection with the third variation of injection compression moulding.

In principle, it is possible to use any standard injection moulding machine to this end with these alterations in controls. Most machine manufacturers offer these additional controls as an option on standard moulding machines. In most cases, the older machines, too, can be retrofitted with the relevant controls.

The injection moulding machines with electro-mechanical drives show an edge over the hydraulic machines in consistency as oil is compressible and changes its characteristics with temperature variations.

During the compression phase, the melt can be forced back to the plasticising cylinder. It is therefore essential that the gate should have frozen before commencement of compression. The machine nozzle too should be of the positive shut-off type.

THE MOULD

The injection mould, as described above and shown in Fig. 7.1, closes in two steps. Even in the partially open state during injection, it must not let the melt go out of it. It implies that it has to be of a positive type like the compression moulds for thermosets. This adds to the complicacy of the tool in case of articles with irregular contours.

Instead of one part of the mould dipping in the other, a spring-loaded frame incorporated in one half of the mould (Fig. 7.3) can act as the outermost border. The mould becomes simpler as the two mould halves can have a plain parting face. The force acting on the frame cannot be adjusted unless the springs are changed. The frame can also be moved hydraulically through the core pulling controls in the machine.

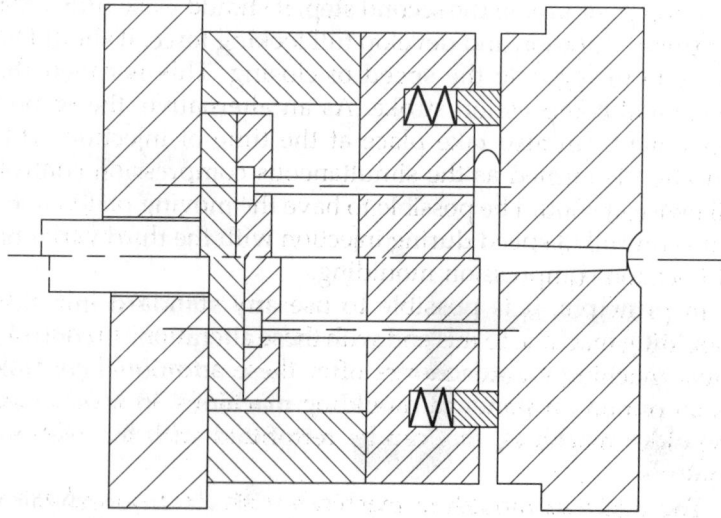

Fig. 7.3

It enables easy adjustment of the pre-tension but makes the mould more expensive.

It is also possible to integrate hydraulic cylinders in the mould itself for the secondary movement as illustrated in Fig. 7.2. The lower cores of the mould may be fixed on a common ejector plate. A hydraulic cylinder, operated by the core pulling device of the machine, pushes up the cores in the partially filled, closed mould and compresses a section of mouldings to their final thickness. This makes the mould a self contained unit which can run on any standard moulding machines having a core pulling device. The hydraulic cylinder is required to provide the compression force for all cavities. The pressure required depends upon the material and its viscosity. In case of PMMA, which is often used to mould lenses, the compression pressure may come to 700 Kp. per square centimeter.

The major application of the process and consequently the moulds is in the field of lenses, glasses, prisms and other accurate articles like gears as well as for moulding thin walled products such as compact discs, housings for mobile telephones, calculators, laptops and other articles combining light weight with accuracy. An added advantage is the surface finish which is a faithful reproduction of the cavity surface details. Any cores forming through holes and apertures have to dip into the opposite side of the mould.

Moulds for injection compression moulding need accurate matching of the two halves or of the sliding components as the case may be. The dipping parts or components are subjected to very high wear and tear. It calls for wear resistant steels for the mould inserts which must also resist bending. The mould components must be sturdy as any bending can aggravate the problem of wear and lead to inaccuracies in the wall thickness of the moulding. It may be pointed out that the stationary half of the mould is not supported by the other mould half during injection (except in the selective injection compression moulding). It also has to withstand the lay on pressure of the machine nozzle. This calls for adequate stiffness of the stationary mould half as well as that of the clamping arrangement. Uneven cooling can also contribute to variation in dimensions of the matching components and to distortion and higher wear. It reflects adversely in the quality of the moulding.

Another prerequisite of the form-shaping mould components is a very fine and accurate finish which gets transferred in the minutest details on the moulding. Alloy steels with maximum purity, hardened to 54–58 Rc, fill the bill. Hard chromium plating is not recommendable as it may peel off due to constant friction. Some stainless steels too have proved suitable.

The location of sprue or gate should ensure even filling of the mould cavity. The mould filling programmes can prove helpful in choosing gate location and predicting pressures.

It is not possible to incorporate slides and other such conventional components to form undercuts. As the melt has to be forced to remote regions through compression, the process works well with relatively flat articles. General product design guidelines such as those for ribs, bosses, projections, etc. are applicable here too.

ADVANTAGES OF INJECTION COMPRESSION MOULDING

The process enables thin wall moulding with low pressure and low clamping force. Larger articles can be produced on relatively smaller machines.

The lower moulding pressure results in lower stresses. The shrinkage is lower as well as more uniform because of the uniform compression pressure and consequently the warpage is minimal.

The higher density achieved in mouldings out of semi crystalline thermoplastics improves toughness and wear resistance.

The cavity surface details are reproduced faithfully.

The optical components like lenses can be produced more accurately and economically without post-operations. Concave lenses, thin in the middle, which are problematic to produce by conventional injection moulding, can be manufactured accurately by compression injection.

The orientation of fibres in filled plastics is significantly lower, which fact ensures more uniform strength in the article.

A major drawback of the process is its inability to compensate the volumetric contraction in ribs, bosses and other such functional details upon cooling. Sink marks may be unavoidable at these locations.

FIELDS OF APPLICATION

Accurate optical components like lenses and prisms, large thin-walled mouldings such as membranes for loud speakers, car panes, housings of light weight appliances, mobile phones, pocket radios and portable CD-players, laptops, Digital Versatile Discs (DVD) etc.

The injection compression process is also employed for moulding of thermosets and elastomers. Another field of application is the moulding of inlay foils and textiles.

8

Multi-component Injection Moulding

- The process
- The mould
- Alternative processes
- Applications
- The equipment
- The product design
- Advantages

THE PROCESS

Multi-component moulding, also called overmoulding, is the process of injection moulding a component successively in more than one steps, using different material each time, till it has achieved its final shape. The plastics injected in successive steps may be the same but different in colour as in the case of typewriter keys, radio knobs and countless such components or it may be entirely different in its chemical composition and properties such as a hard plug with a soft seal. Irrespective of the number of materials or colours, the basic feature of the process is that each material has its separate path to the mould cavity. This differentiates it from other multi-component processes involving more than one plastics for a product.

The process can'be readily visualised by considering the case of a two-colour typewriter key as an example. The letter on the key is required to be in a differently coloured material, say black, than the rest of the component which may be white. It means that the product is made up of two components. The basic component or the substrate may have the raised letter and a part of the inner structure and be in black material and the shell surrounding the letter and forming the body may consist of the material having white colour of the key. Now if the substrate is moulded separately and transferred along with the mould core to another mould having a cavity with the outer shape of the shell, the final moulding will be a key with a black letter surrounded by a white shell.

The process can be carried out automatically in one mould with two sets of cavities. One group corresponds to the external shape of the substrate and the other group with an equal number has the shape of the outer shell. The two sets of cavities are placed equidistant on either side of the mould centre line. The cores for both sets are identical. The injection moulding machine has two injection units. The additional unit may be placed at right angles to the horizontal axis of the machine, either vertically from above or horizontally from the back so that it can inject at the parting line of the mould.

As the mould opens after the first injection, viz. that of the substrates, the moving half of the mould is rotated by 180 degrees, thereby bringing the cores with moulded substrates opposite the cavities of the shell and the empty ones in line with the cavities of the substrate. By the next moulding shot, the first injection unit moulds a fresh set of substrates and the second unit injects a shell over the substrates moulded in the previous cycle and transferred to the second station. As the mould opens again, The moulding from the second station is ejected and the core plate is turned again. Thus every cycle delivers a group of finished keys moulded in two materials. Figures 8.1A and B illustrate two steps of the process for a single cavity mould.

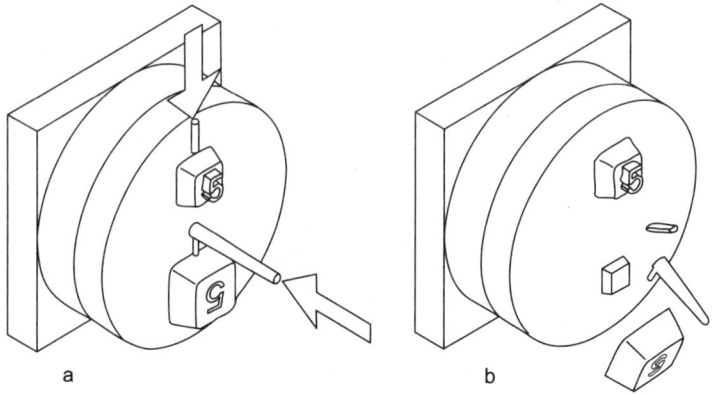

a   b

Fig. 8.1

An article can be moulded in any number of materials with corresponding number of injection units in the machine and as many groups of cavities in the mould. So far, articles in six materials have been moulded. The plastics may be different but must be compatible. They should have natural adhesion with one another. In some cases, it may be possible to provide mechanical anchorage through undercuts but generally, it is difficult if not impossible. In any case, the mould becomes very complicated in construction and big in size.

It is also possible to mould thermoplastics with elastomers and thermosets.

THE EQUIPMENT

The basic moulding machine is a conventional one. It has to have additional injection units corresponding to the number and type of materials to be moulded. For example, the injection unit for a thermosetting material will obviously be different from that for thermoplastics. Their parameters can be set independently.

The special machines have a built in turning mechanism on the moving side for indexing the moving half of the mould. It is also possible to fix a special plate with built-in turning mechanism on the moving platen. (Fig. 8.2). The turn table also

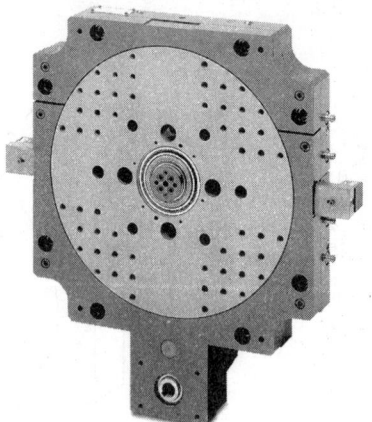

Fig. 8.2

houses the cooling, pneumatic and hydraulic lines for the mould. The turning of the mould platen is carried out by a built-in hydraulic motor.

THE MOULD

The injection mould will have as many sets of different cavities as the number of materials to be moulded. The number of cores corresponds to the total number of cavities. All cores, however, are identical in design.

With each moulding cycle, a set of finished articles is produced. The mould must be equipped with a selective ejection arrangements which demould only the finished group and not the substrates. It may be actuated by means of a small hydraulic cylinder built in the mould.

Although it is possible to use conventional cold runners, the preferred system of feeding is through hot runners. The cooling of the revolving half of the mould is carried out through

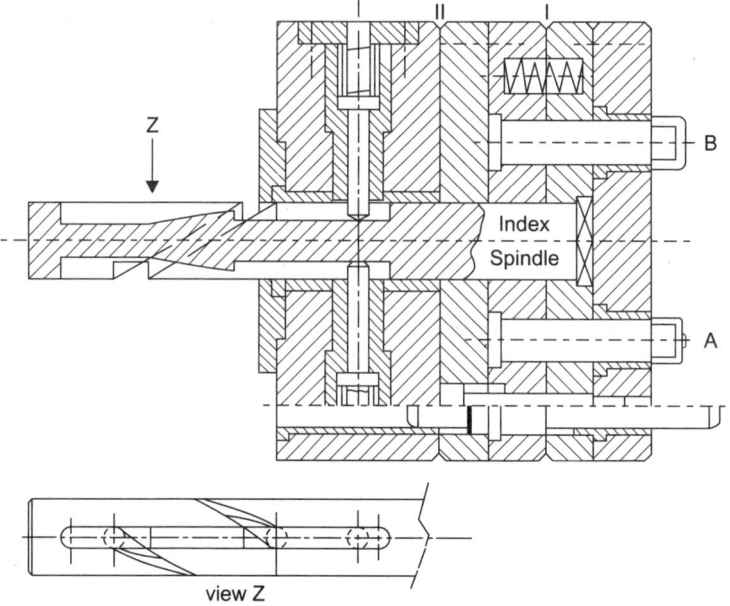

view Z

Fig. 8.3

connection with the water circulation system built in the revolving machine platen.

If the moulding machine does not possess an indexing mechanism, it has to be incorporated in the mould. In that case, not the whole mould half but only the core carrying plate is indexed. The rotation may be imparted mechanically, hydraulically or electrically. It adds to the complicacy of the mould. A crucial feature of the tooling is the extreme accuracy in matching of cores and cavities in shape as well as in alignment. The smallest deviation may result in flash and blurring of meeting lines of different colours.

Mechanical Indexing System

Figure 8.3 depicts the moving half of a 2-component injection mould with built-in mechanical indexing mechanism. The central spindle, attached with the core holding cum ejector plate assembly, bears two longitudinal grooves placed opposite to each other. Two spring-loaded pins, housed in the back plate of the mould, grip the spindle in the grooves and control its rotation when it is moved forward. The spindle performs the following functions:

a. It pushes the set of stripper plates forward and ejects the finished mouldings.
b. It imparts 180 degree rotation to the core plate assembly as well as the stripper plates.
c. It moves back straight without turning when the mould closes.

The path of the grooves (view z), which is primarily straight but partly in the form of a helix covering half the circumference of the spindle and opening into the opposite groove, governs the axial and the rotational movement of the spindle. Seen from right to left, the groove runs straight for certain length (stripping stroke) at the end of which a helical section starts and makes half a turn. Simultaneously, the groove continues straight along the axis. Its depth varies in different sections. The first straight section, the helical section and the last straight path have different depths. The depth of the middle straight section changes gradually as can be seen in the sectional view.

As the mould opens and the machine ejector system is actuated, it pushes the spindle which moves straight for certain distance.

The subsequent part of the ejection stroke is converted by the helical groove into a 180 degree rotation of the spindle and also of the core plate assembly fixed with it. The core with the substrate moulding (A) gets positioned opposite the outer shell cavity and the empty core faces cavity for the substrate. The spring-loaded pins are now in the last and the deepest section of the grooves.

When the mould closes, the stripper plates, etc. are pushed back along with the spindle. The spring-loaded pins cannot get back into helical section of the groove because of the difference in its depth. There is no rotation and the pins travel (relatively speaking) in straight line over the middle part of the groove, to the front section.

The mechanical system has the advantage that it forms an integral part of the mould and does not necessitate any additional arrangement or accessory in the moulding machine. However, it has to be designed and manufactured afresh for every article.

The Hydraulic System

The basic principle, viz. moulding the article in two steps remains unchanged in this version also. The task of turning the core carrying plate is, however, accomplished hydraulically. The central spindle is provided with pinion and a rack, powered by a double acting hydraulic cylinder, turns the spindle. The rack is pushed forward and pulled backwards in successive shots unlike with unscrewing moulds.

The turning movement can also be generated with the help of worm and gear actuated by an electric motor.

THE PRODUCT DESIGN

A plastics article made of more than one material has to be conceived from the point of view of the moulding sequence as well as from the basic angle of its function. The final product consists of a number of layers or segments of different materials.

Each substrate has to provide the possibility of flow and of firm anchorage to the next one.

The degree of adhesion between the bonding surfaces is dependent upon:
- The material combination
- The part geometry
- The process control

Mechanical anchorage alone may often be found inadequate. It is imperative that there should also be chemical interaction, viz. the free molecular movement between the two mating surfaces. The important parameters for this are the temperatures of the contact surfaces and the crystallisation temperature or the glass transition temperatures of the material pair. Additives and fillers, too, influence the strength of the bond. Some specific additives called compatibilizers can improve the adhesion in certain cases. The mould shrinkage values of the material pair should not be too far apart.

In case of pairing different materials for subsequent layers, compatibility must be given due consideration. Table 8.1 shows the suitability of combination of various materials. Please note:
- PA 6 & PA 66 are identical in behaviour except with PC. PA 6 bonds with PC but PA 66 does not.
- PPO is not compatible with any of the plastics listed in the table.

It is also possible to mould thermosets as well as elastomers with thermoplastics in multi-component moulding process The most crucial point in these combinations is that the thermoplastics should be able to withstand the mould temperatures required for processing of the thermosetting polymers. The few thermoplastics, which come in question, are polyamides, polyacetal, polycarbonate, polyphenylene sulphide, polyesters, etc. The liquid silicone rubbers have to be modified to bond with these thermoplastics. The basic conditions for moulding thermoplastics with thermosetting elastomers are:
a. In case of a combination of hard material (thermoplastics) and a thermosetting/vulcanising elastomer, the hard material must retain its shape and rigidity at the curing/vulcanising temperature of the elastomer till ejection.

Table 8.1: Adhesive bonding polymers

	ABS	CA	EVA	PA 6	PC	PE-HD	PE-LD	PMMA	POM	PP	PS	SB	PBTP	TPU	PVC-S	SAN	PETP	PSU	PC-PETP	PC-ABS
PC-ABS	+	×	×	+	+	-	-	+	-	-	-	-	+	+	+	+	+	+	+	+
PC-PETP	+	×	×	+	+	-	-	+	-	-	-	-	+	+	+	+	+	+	+	+
PSU	+	×	×	×	+	-	-	×	×	-	-	-	+	+	×	×	+	+	+	+
PETP	+	×	×	×	+	-	-	-	-	×	-	-	+	+	×	×	+	+	+	+
SAN	+	+	+	+	+	-	-	+	-	-	-	-	+	+	+	+	×	×	+	+
PVC-S	+	+	-	×	×	-	-	+	×	-	-	-	+	+	+	+	×	×	+	+
TPU	+	+	×	+	+	-	-	×	×	-	-	-	+	+	+	+	×	×	+	+
PBTP	+	+	×	+	+	-	-	-	-	-	-	-	+	+	+	+	+	+	+	+
SB	0	-	+	-	-	-	-	-	-	-	+	+	-	-	-	-	-	-	-	-
PS	0	-	+	-	-	-	0	-	-	-	+	+	-	-	-	-	-	-	-	-
PP	-	-	+	0	-	-	+	0	-	+	-	-	-	-	-	-	×	-	-	-
POM	-	-	×	-	-	0	0	×	+	-	-	-	-	×	×	-	-	×	-	-
PMMA	+	×	×	×	×	0	0	+	×	0	-	-	-	×	-	+	-	×	-	-
PE-LD	-	-	+	0	-	+	+	0	0	+	0	-	-	-	-	-	-	-	-	-
PE-HD	-	-	+	0	-	+	+	0	0	0	-	-	-	-	-	-	-	-	-	-
PC	+	×	×	+	+	-	-	×	-	-	-	-	+	+	×	+	+	+	+	+
PA 6	+	×	×	+	+	0	0	×	-	0	-	-	+	+	×	+	×	×	+	+
EVA	+	0	-	×	×	-	-	×	×	-	-	-	×	×	-	+	×	×	×	×
CA	+	+	0	×	×	-	-	×	-	-	-	-	+	+	+	+	×	×	×	×
ABS	+	+	+	+	+	-	-	+	-	-	0	0	+	+	+	-	+	+	+	+

+ Good bond
- Poor bond
0 No bond
x No test results

b. The mould must have a very effective thermal separation between the heated cavity for elastomer and the relatively colder cavity of the hard thermoplastics and also between the hot runner system for thermoplastics and the cold runner system for the elastomer.

c. The cycle time is usually decided by the curing/vulcanising time of the elastomer.

As mentioned before, the different materials used for subsequent layers must be compatible and should form a bond with each other besides the mechanical joint. Mechanical anchorage may have to be employed for overmoulding mutually incompatible materials. The process of incorporating undercuts in the moulding for holding the subsequent layer, however, adds to the complexity and the cost of the mould.

A positive use of the incompatibility of two materials is made in moulding multi-component products with moveable parts, such as hinges and toy figures with movable limbs.

Of course, the other guidelines for design of the moulded products concerning wall thickness, material accumulation, radii, ribs and shrinkage, etc. apply here also.

ALTERNATIVE PROCESSES FOR MULTI-COMPONENT MOULDING
Substrate Transfer by Handling Device

A variation of the process consists of employing a handling device or a robot to transfer the moulded substrates from one set of cavities to the other before the next shot. It obviates the indexing mechanism and makes the mould less complicated.

Yet another variation of the method of transferring the substrates with the help of a handling device is to employ a stack mould. The cavities for substrates and final products are placed on separate levels and the robot transfers the moulded substrates from one level to the other where these are overmoulded through the second injection unit.

The advantages of a stack mould are manifold. The whole mould plate area can be utilised for one type of moulding. The pressure distribution is even and the cooling/heating arrangement is more uniform. The separation of two injection systems, especially in case of different types of plastics is unproblematic. The mould cost is, of course, higher but is well compensated by higher productivity.

Sliding Split System

Another method of multi-component injection moulding, applicable in special cases, is to block the second material sections of the mould cavity with slides, mould it in one material and then remove the barriers without opening the mould. Now the second component is injected. It is possible to inject further materials successively after removing the blockage. The system, introduced by Battenfeld of Germany, is also called the **combiform process**.

The injection of different materials takes place in sequence and not simultaneously as in the rotating mould technique described above. Though the method is slow, it may prove economical for smaller production batches. As there is no need to transfer parts or rotate mould plates, the tool cost is much lower than in the previous method. Also the more compact mould construction allows the use of a smaller moulding machine. The method, however, is not universally applicable. It is limited to special product configurations which permit blocking of sections of the moulding which will be moulded in another material subsequently.

Swivel Plate System

A special machine for two component moulding, developed by Engel of Austria, employs the moving platen side for the second injection. An injection unit is situated above it. The fixed platen carries the mould half with cavities for the substrate in the first material whereas the moving platen has the mould half with equal number of cavities for the shell in second material. The positions of cavities are, of course, identical. The special feature of the machine is a swivelling plate, positioned between the two platen on the same axis. It can revolve around its own vertical axis by 180 degrees and can slide horizontally on linear guides on the axis of the moulding machine. The swivelling plate carries two identical mould halves for cores on two opposite faces. It also houses the ejection mechanism.

In closed position of the machine, there is a complete mould for the substrates on the fixed platen side and another complete mould for the shell on the moving platen side.

After the substrate is moulded, the machine opens, the swivel plate slides to the middle and revolves by 180 degrees to bring the moulded substrates opposite the cavities for the shell in the mould half on moving platen. The machine closes and both injection units inject. Now on the moving platen side, the shell has been moulded over the substrate and a set of substrates has been moulded on the fixed platen side. Each shot yields a set of finished article. Figure 8.4 shows the machine in open condition with the substrate and the finished article.

Whereas in the first system of core plate rotation, the mould contained two sets of cavities differing in size and subject to different pressures, the pressure here is distributed evenly as the mould cavities are placed symmetrically all over the mould plate. There is more space for cooling hoses in a swivel plate mould which results in faster cycles and higher output. The moulds are simpler and therefore cheaper as they do not have any moving or rotating components.

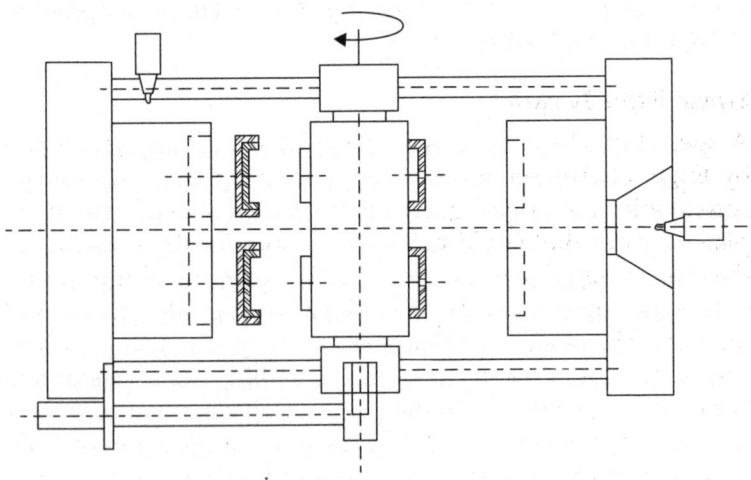

Fig. 8.4

The system is particularly advantageous for large products. The injection in the middle obviates the need for complicated hot runner systems.

In another variation of the swivel plate method, the second injection unit is placed behind the moving platen, so that the second injection is either on the parting line or through a hot runner system housed in the core plate of the mould-half on moving side, which will also house the ejection system.

Swivel Cube System (Fig. 8.5)

The swivel cube system may be regarded as a further development of the swivel plate technology. A cube, capable of rotating by 90 degrees about its own vertical axis takes the place of the swivel plate. Here too, the fixed platen carries the mould cavity half for the substrate and the moving platen the mould cavity half for the shell. The rotating block is equipped with four identical mould halves for the cores, one on each face. The first injection is carried out in conventional manner through the fixed platen and the second one by means of an additional injection unit placed in L configuration to inject at the parting line. The cube contains lines for water and air and also houses the ejection mechanism.

The first 90 degree revolution of the cube after moulding brings the substrates to a free station where it can be loaded with inserts, etc. The second revolution after the moulding shot brings the same core mould opposite the shell cavities of the moving platen and the third revolution again to a free station

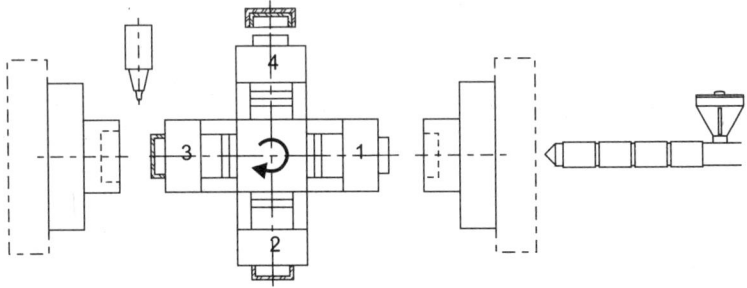

Fig. 8.5

where the doubly moulded finished article can be ejected. The cycle is a continuous one and every revolution yields a set of finished mouldings.

ADVANTAGES OF MULTI-COMPONENT MOULDING

- Properties of various plastics can be combined.
- Moulding of individual parts, their assembly and other after-operations are eliminated.
- Intermediate storage and inspection of individual components is obviated.
- Only one moulding machine is needed.

FIELDS OF APPLICATION

- Keys for type writers, computers, telephones, etc.
- Buttons and knobs for electronic, industrial and household. gadgets, automobiles, etc.
- Components with integrated scripts and symbols.
- Covers with transparent windows.
- Multi-coloured covers for automobile lights.
- Decorative publicity articles.
- Machine components with integral gaskets.
- Closures with seals.
- Toys with moveable parts.
- Technical components like hinges, air ducts.

9

Sandwich Injection Moulding

- The process
- The mould
- Faults and remedies
- Advantages
- The equipment
- The product design
- Alternative processes
- Applications

THE PROCESS

Sandwich moulding, also called co-injection moulding, is the method of moulding a component out of two (or more) materials whereby one of the materials forms the outer skin or the sheath and the other material constitutes the core. The aim is to combine properties of two or more materials, such as high quality visual appearance with a core out of recycled or filled material. It may also be a combination of transparent outer layer and a coloured core or a soft external sheath with rigid core. In fact, the process offers possibilities of unlimited combinations for specific applications. The prime condition for a successful moulding is the good bonding of materials. Table 8.1 (Chapter 8) shows various material combinations and their compatibility.

The difference between the two-component moulding and the sandwich moulding is that in the former process, the second material is either moulded over the first or besides it whereas the second component in sandwich moulding is injected into the first one to be entirely enclosed by it.

Sandwich moulding is carried out with an injection moulding machine having as many injection units as the number of plastics to be moulded. It may be remarked that it is mostly practised with two materials or the same material in two forms such as virgin material for the skin and reground material for the core or transparent (or uncoloured) material for outer layer and coloured one for the core. Consequently, in most cases it is a standard moulding machine with an additional injection unit placed vertically or from the back in horizontal position. A special co-injection nozzle with separate paths for melts from the two injection units mounted on the standard machine permits injection of melts in turns (Fig. 9.1).

The melt forming the sheath (coming from the outer channel) is injected first, filling the mould partially while the channel for core material is blocked. Now the bush X is advanced by means of a lever operated by a small hydraulic cylinder (not shown). It blocks the melt channel for the sheath material and opens the path for the core material. Now the second melt is injected through the common sprue in the still plastic interior of the first material, the outer layer of which has solidified along the mould cavity wall. The in-coming melt pushes the still fluid core of the first material further and outwards to the cavity wall in the manner of a fountain flow, thus filling the cavity completely.

The central channel is closed by pulling back the bush X. It opens the path for the sheath plastics which is injected once again. The purpose is to create a seal in the sprue area to obliterate the mark left by the second melt. The sheath gets again an uninterrupted appearance and the nozzle contains again the right material for the next cycle.

It may be remarked that between the switchover of channels, there is a short phase. when both materials flow

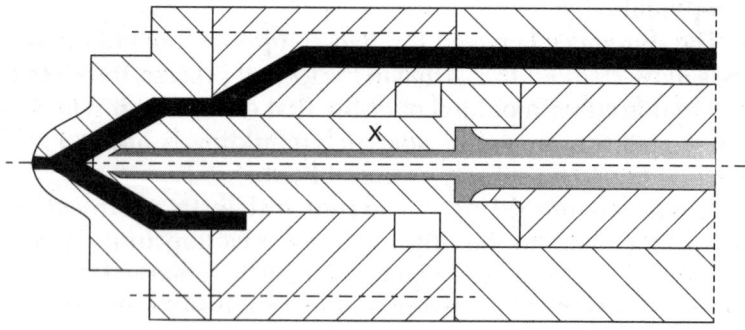

Fig. 9.1

simultaneously. This is necessary to avoid a break in the pressure and any marks caused by the halt of the sheath material.

The core material constitutes 40–75 % of the total volume. It may also contain foaming agents.

Sandwich moulding is based on the characteristic of "fountain flow" of viscoelastic melts. During injection, the outer layer of the melt touches the mould wall, cools down and becomes stationary. On further injection, the plastic melt in the core moves forward, swells and touches the walls and stays put. Now, if the second material is injected after the first material has filled the mould only partially, it would push the still fluid core of the first material forward and outwards. In this way, a sheath would be formed out of the first material gradually till the end and the second material would occupy the place of the first in the core.

The thickness of sheath depends upon:
• Volume of the first material
• Melt and mould temperature
• Viscosity of the polymers
• Injection and holding pressure
• Injection speed
• Cooling design of mould
• Product design

The filling pattern can be studied by injecting the first material in transparent version (possible with amorphous polymers) and then its coloured version or by moulding natural polypropylene first and its coloured version after that.

As a general rule, the viscosity of the sheath material should be lower than that of the core plastics. This step facilitates more uniform distribution.

THE EQUIPMENT

The main part of the equipment is a standard injection moulding machine, fitted with a special nozzle. A second injection unit is also docked with the same nozzle. The second injection unit with its own controls may be mounted vertically or horizontally at right angle to the first one. Controls of both injection units are entirely independent although these are interconnected to coordinate the start and termination of injection. The specially designed nozzle houses two separate melt channels, independent of each other. An inbuilt mechanism moves the valve component and opens one channel blocking the other one simultaneously so that at one time, melt from only one unit can be injected. It is, however, also possible to let both materials flow simultaneously. This is done towards the fag end of injection of the first material so that there is no break in the flow. The mechanism to switch over the channels is operated hydraulically. Figure 9.1 (based on Engel-nozzle design) depicts the working principle.

The closing side of the machine does not differ from that of a standard injection moulding machine.

THE MOULD

The mould is essentially a cold runner type. In all other respects, it corresponds to a standard injection mould.

THE PRODUCT DESIGN

The article is required to have certain minimum wall thickness so that the core of the first material remains plastic till the second melt is injected. 4 mm. may be taken as the absolute minimum for most material combinations.

Parts of the cavity, which are likely to get filled first and form blind alleys, such as thin lugs or side arms and which cannot be flowed through, will not have the second material in the core. Very thin or narrow side details may be filled only partially by the first melt and form a skin on the front. The second material may break the skin and fill the rest of the side cavity. As a rule, details forming dead ends should be avoided as far as possible.

The thickness of the sheath depends, among other factors, upon the product design. Uniform wall thickness also guarantees a uniformly thick sheath. The core material tends to follow the thicker sections if the article combines sections of varying wall thickness. It is also termed as finger effect.

Add flow aids for the core material to reach far corners after the first trial.

All other rules governing the injection moulded articles are applicable here too.

FAULTS AND REMEDIES

The most common fault encountered in sandwich mouldings is a blemish on the surface at the spot where the sheath material had come to a halt and had partly frozen before the core material was injected. The core of the first material, still in plastic state, is pushed by the second melt and forms the rest of the outer surface. The joint may be found visually disturbing.

The fault may be eliminated by starting injection of the second material before the end of the first injection. In absence of that facility, the blemish can be camouflaged with a structured mould surface or with decorative foils.

The switchover from the first to the second material, howsoever fast, causes slowing of the flow front. It may appear as a shadow or as a minute notch in extreme cases. It can be remedied to a great extent by optimising the process of switchover.

Gas bubbles may appear on the surface of the moulding as silver streaks, particularly towards the end of the flow path. There are two causes giving rise to this blemish. If the melt has too long a dwell time in the plasticising unit, some additive chemicals get overheated and release gases. It is advisable to delay re-charging towards the end of the cooling time so that the melt does not have a long waiting time before injection. Slow injection also gives time to bubbles to appear on the surface and get elongated through friction. A fast injection remedies the defect to a great extent.

Distortion in the moulding is the result of non-uniform distribution of the core material. It is caused by improper product design and unfavourable viscosity combination of the two plastics. Too low a viscosity of the core material brings instability in its flow pattern caused by differences in resistance to the flow. The core melt flows more in sections offering less resistance, neglecting thereby less favourable paths. Consequently, the core thickness will not be uniform. Some sections may not have any core at all. As mentioned earlier, it is termed as the finger effect. Unbalanced mould cooling layout may also give rise to similar defect. Sink marks are also to be attributed to the same cause.

The sandwich moulding process is offered by various moulding machine manufacturers. Their main difference lies in the design of the multi-channel nozzle.

ALTERNATIVE METHODS OF SANDWICH MOULDING
Mono-Sandwich Moulding

Ferromatic of Germany has developed and patented a so-called mono-sandwich process which differs from the foregoing one in principle and equipment.

Here too, two separate plasticising units deliver the two melts. The horizontally placed injection unit of a standard injection moulding machine plasticises the material forming the core. A vertical extruder, coupled to the main injection unit

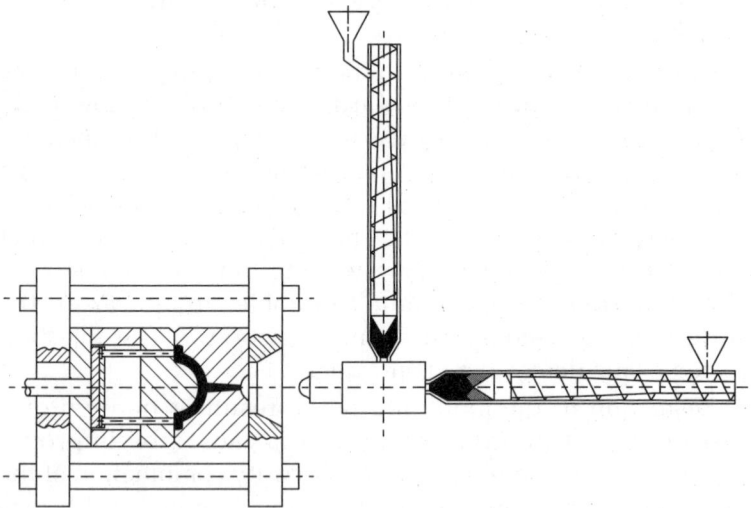

Fig. 9.2

at the head before the nozzle via a connecting block, forces the molten skin plastics into the main injection unit, pushing back the plasticised core material. Now the barrel of the main unit has both melts in a row. The connecting block also acts as the machine nozzle. As the horizontal injection unit starts injection, the channel connecting the two units is blocked and the path to the mould opened. The material in front, viz. the sheath plastics delivered by the vertical extruder, forms the forerunner, followed immediately by the core melt which forces the first material as a thin, uniform layer against the mould walls. Figure 9.2 illustrates the successive stages of the process.

As obvious, the gate would leave a mark showing the core material. It is, however, of no consequence if the feeding point is located at an invisible spot.

The main advantage of the process is the comparative simplicity of the equipment. As the injection takes place without a break, there are no blemishes on the surface due to hesitation.

The mould does not differ in any way from that for the first version.

Sandwich Moulding with a Single Injection Unit

As shown in the mono-sandwich process, two melts placed one after the other in the barrel, do not mix during injection. The same perception has been made use of by plasticising two polymers in sequence in the same plasticing unit and achieving the same effect.

The equipment consists of a conventional injection moulding machine in all features. It is equipped with two hoppers which contain two different plastics raw materials (Fig. 9.3). The feeding to the barrel takes place in turns and is directed electronically. The sheath polymer (S) is fed first in exactly measured quantity, followed by the core material (C) in similar manner. The sheath material for sealing the sprue is fed again. The plasticised melt collects in the front part of the barrel in the sequence S-C-S. The injection of the melt brings forth a sandwich moulding as that with the equipment discussed in the outset.

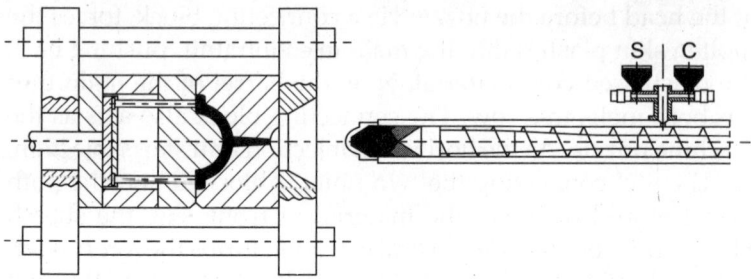

Fig. 9.3

ADVANTAGES OF SANDWICH MOULDING

- Cost reduction by using cheaper material/reground material for the core.
- Weight reduction through foaming of the core.
- Cycle time reduction through weight reduction.
- Elimination of sink marks with foamed core.
- Combination of properties of two materials.
- Enhanced aesthetic appeal through transparent sheath and coloured core.

APPLICATIONS

- Furniture parts.
- Outdoor furniture with UV-shield.
- Household goods.
- Sanitary ware like toilet seat, water cistern.
- Housings for electronic goods with EMI shield.
- Housings for office equipment.
- Automobile components like dash board parts, bumpers, fenders, spoilers.
- Massive grips and handles.

10

Fusible Core Injection Moulding

- *The process*
- *The equipment*
- *Advantages and disadvantages*
- *The materials*
- *The moulds*
- *Applications*

Injection moulding with fusible core or lost core is identical with the moulding process where a core to form an internal detail such as a hole, internal threads or a channel in the moulding is laid in the injection mould before closing it, injecting the melt and then ejecting the moulding after cooling along with the core, which is extracted out of it outside.

The difference in case of the fusible core is that it is used for forming complex internal details which defy extraction of the core without damage to the mould and the product. The only way to remove the core is by destroying it. The classical case is the air-intake manifold of an automobile engine. It had been previously fabricated out of two sheet metal halves and then joined together. It may be claimed that the lost core moulding process has been specifically developed for this component after it was found that it could be moulded out of certain plastics resisting high temperatures. Glass filled polyamides 6 as well as 66 and polyester are the materials being used now.

THE MOULDING PROCESS

The moulding process consists of placing a preheated core, made of a low melting temperature alloy corresponding to the shape of the ducts it is intended to form, in the mould and then performing the standard injection moulding operation.

The fusible core is made of a Bi-Sn alloy with a melting temperature of 138°C. The core is generally in one piece but for very complex shapes, it may be assembled out of more than one components. The alloy has a high specific heat and very good thermal conductivity. Though the melt coming in contact with it may have a temperature above 260°C, the core does not melt because of its excellent thermal conductivity and high specific heat. The poor thermal conductivity of the frozen plastics layer over it provides it a protective sheath against the still molten material behind it. The alloy has very low tensile strength and the core may bend or break under the direct pressure of injected material. It is, therefore, essential that the feeding system is designed in such a way that the incoming melt flows along it and not impinge on it directly. It may call for multiple gating. For complex articles, the gating is determined with the help of mould flow programmes.

The core is ejected along with the product and fused out in a bath, hot enough to melt the alloy but not the polymer. The molten alloy can be collected and, after cleaning, reused.

The last step in the production cycle is washing and drying of the product.

The process is finding application also in other fields besides the automobile sector. Some plastics require cores out of alloys with higher melting points.

The continuous process consists of the following steps:
1. Cast the fusible core.
2. Assemble the core if it consists of more parts.
3. Place the core in the injection mould and overmould with plastics.
4. Eject the moulding along with the core.
5. Melt out the core in the melting bath.
6. Remove the moulding from the bath, clean and dry it.
7. Recover the alloy from the bath and return it to the stores.

THE MATERIALS

The thermoplastics commonly employed are the ones which can withstand high temperatures. Apart from glass filled polyamides 6 and 66, PBT is also being used for specific applications. In view of the low tensile strength of the core alloy, the flowability of plastics may be enhanced by means of certain additives in order to inject them at lower pressures.

The low melting temperature alloy for casting of fusible core consists of bismuth (58%) and tin (42%). Its melting temperature is 138°C. The melting temperature can be manipulated by changing proportion of the ingredients.

The medium to melt out the core is modified polyglycol ether which can be safely heated up to 160–170°C.

Water with some detergent is used for cleaning the mouldings after the core has been melted out.

THE EQUIPMENT

The Injection Moulding Machine

The moulding machine for the process is in no way different than the conventional one. It is equipped with a shut-off nozzle if polyamide is being processed. As the plastics moulded is usually of the glass filled variety, the machine should preferably have bi-metal screw and barrel for a long life. The operations of taking out a moulded part and laying in a fresh core is usually performed by a robot in mass production. It may need greater opening of the mould closing unit than the normal one.

Fusible cores may weigh around 80 kilogramme or more in some cases. It may not be convenient to place and secure them in the mould on a horizontal injection moulding machine. Especially for large production batches, a sliding table or a revolving table vertical machine is found more suitable. The hot cores are liable to bend under their own weight. The handling devices for transferring them to the mould must support them at all points.

After the moulding process, the moulding containing the core is taken immediately to a hot bath for removing the core by melting. The heat contained in the core helps reduce the melting time.

The Melting Bath

It consists of a rectangular tank with heating arrangement to melt out the core. The fluid employed for transferring heat to the core is a modified polyglycol ether which is quite stable till about 160–170°C. It does not react with the Bi-Sn alloy but undergoes a reaction with iron (some assembled cores may have iron parts). Consequently, the tank is fabricated out of stainless steel sheet. The time needed to melt the core depends upon its weight. It is a slow process. The tank is chosen big enough to accommodate a large number of mouldings so that with every cycle, when a new moulding is put in, at least one in the tank has become free of the core and can be taken out. The salvaged core material can be reused.

Another method of melting out the core is the high frequency induction heating. It is fast but more expensive as the heater has to match the shape of the product. Induction heating is not suitable if the moulding contains metallic inserts.

The Washing Equipment

After the last operation of core removal, the mouldings contain traces of the melting medium which have to be removed before the moulding is put to its end use. It is done with water and suitable detergents in simple washing machines. Water can be recovered by methods such as vacuum distillation.

With glycol ether as the metal core melting medium, no additional cleaning chemicals are required with water. Cleaning of the mouldings is achieved by dipping in successive water baths.

The cores may be procured from vendors but for mass production, it is economical to cast them in own shop parallel to the moulding section. An advantage of this arrangement is that the freshly cast core need not be reheated to 80°C before loading in the mould. The casting machine forms part of the manufacturing equipment for production of the lost core mouldings.

The Equipment for Core Casting

The flowability of the molten Bi-Sn alloy is a million times higher than that of the average polymer. It needs very little pressure for filling the mould. The process resembles injection moulding with the difference that the alloy is melted outside in a separate tank and conveyed, through a gear pump, to the mould held in the mould closing unit of a moulding machine, by means of metal pipes which are heated and insulated. However, unlike in injection moulding, the filling of the mould is not central as in conventional single cavity injection moulds for plastics but from below to avoid formation of bubbles on the surface of the casting. The solidified casting is transferred from the mould to the injection moulding station by a robot as mentioned before.

THE MOULDS

The Injection Mould

The mould for overmoulding of the fusible core has no special features other than the conventional ones for glass-filled polymers. The multiple gating may necessitate a hot runner feeding system with shut-off nozzles to prevent drooling. The flow of the melt should be along the core, so that the core is supported by the sheath formed by the advancing plastics. A mould flow analysis prior to the design is recommendable.

The mould temperature should be maintained as recommended for the particular plastics.

The Mould for the Fusible Core

The mould consists of two halves with ejection mechanism like an injection mould for thermoplastics. The injection, however, takes place from below under low pressure. The alloy has very low contraction which obviates a holding pressure but the filling pressure has to be kept up till the gate is frozen.

The mould is maintained at 70–80°C by circulation of hot medium in a network of holes or channels in the core and cavity plates or through electrical heaters. The tool is not subjected to high pressures and can, therefore, be safely fabricated out of non-ferrous alloys which do not react with the alloy of bismuth and tin. Anodised aluminium has proved suitable for the application with the added advantage of better heat conductivity resulting in shorter moulding cycles.

The ejection arrangement should ensure a balanced demoulding without distortion as the bismuth alloy has a very low tensile strength.

Note: In principle, it is possible to combine the two manufacturing steps, viz. core casting and overmoulding in one mould like in the two component moulding. Special custom made equipment, combining the two processes, may prove profitable for very large quantities.

ADVANTAGES AND DISADVANTAGES

Advantages

The fusible core moulding process has the following advantages:

- Freedom of product design. No compromises need be made to extract the core.
- Compact moulding without a joint.
- Elimination of the joining operation of two component halves.
- Accuracy of external and internal dimensions.
- Good surface finish both outside as well as inside.
- Injection mould uncomplicated.
- The Bi-Sn alloy can be used again and again.

Disadvantage

- Very high investment in equipment.

APPLICATIONS

Automobile components, housings for pumps, water distribution heads, oil channels, etc.

11

Injection Moulding of Structural Foams

- *The process*
  - *Foaming with chemicals*
  - *The equipment*
  - *The mould*
  - *Advantages and disadvantages*
- *Foaming with fluids*
  - *The equipment*
  - *Alternative equipment for foaming with fluids*
  - *The mould*
  - *Advantages and disadvantages*
- *Product design for structural foam moulding*
- *Applications of structural foam mouldings*

THE PROCESS

Injection moulding of structural foams is a process to mould articles in thermoplastics with a compact outer layer and a foamed interior, resembling a honey comb. In principle, the process is applicable to all thermoplastics. However, plastics most commonly used are PS, ABS, PC, PP and PVC.

There are two variations of the process which differ in execution, equipment and ingredients but deliver similar results.

Foaming with Chemicals

Chemical foaming agents are azo-compounds, sodium bicarbonate and mixtures of organic acids and their derivatives such as the citric acid. They decompose at a certain temperature releasing gases like nitrogen, carbon dioxide, ammonia and steam and some solids. The gases produce a cellular structure in the melt.

Chemical foaming agents are added to the polymer before injection moulding, either in powder form whereby an adhesive agent like glycerine acid is used as a wetting medium or in the form of master batches with an appropriate polymer as carrier with different concentrations so that a desired cellular structure can be achieved. Those with very low concentrations are used primarily to eliminate sink marks in thick walled mouldings. The amount of foaming chemicals may vary between 2 to 5 per cent by weight. They can be added to the polymer pallets like other master batches. The screw of the moulding machine acts also as a mixer. The foaming agent lowers the viscosity of the resultant melt.

The melt, impregnated with the foaming chemical, is injected into the mould, filling it partially, usually up to about 80%. The pressure required is obviously low. It should be less than the pressure developed by gases when the foaming agents evaporate. The outer layer of the melt in contact with the relatively colder mould cavity walls solidifies whereas the gases released due to decompression of the melt in the mould lead to the foaming of the interior thereby increasing its volume and filling up the mould cavity completely. The gas develops pressure of about 15 bar and acts as the source of the holding pressure. No holding pressure need be applied by the moulding machine.

The temperature of different zones of the barrel of the injection unit is set in such a way that the decomposing temperature of foaming chemicals is arrived at after the start of compression zone. This prevents the released gases from escaping through the hopper. In any case, the chemicals must be fully degraded before injection as any undecomposed residue may appear as a blemish on the surface of the product.

The moulding cycles for the foam injection moulding are considerably shorter than for compact injection moulding because less material has to be cooled and the holding time is eliminated. The gases keep the plastics in contact with the cold mould surface till the end. Injection moulds made of light metal alloys dissipate the heat faster and contribute to further reduction in cycle time.

The different phases of the moulding process are:

1. Injection of the plasticised melt into the mould with partial filling.
2. Release of gases, foaming of the melt resulting in complete filling of the cavity.
3. Cooling of the melt with simultaneous holding by gases.
4. Recharging and plasticising for the next shot.
5. Ejection.

The moulded product may be compared to a bone in its structure. The outer surface is smooth whereas the interior is porous. The product possesses higher mechanical strength with less weight as compared to a compact moulding. The final density achieved may be around 60% of the compact polymer. The density is, however, not uniform in the structure of the moulding. Porosity is maximum in the core and decreases gradually towards the outer surface. In other words, the density is maximum on the surface and minimum in the core.

The Injection Moulding Machine

Any conventional moulding machine with a three zone screw would perform the operation satisfactorily. As the injection pressure required to fill the mould cavity of a thick walled article partially is not very high, the locking force of the machine can also be low. In other words, an injection unit with the biggest screw will not only be adequate but would also increase the output. Alternatively, the moulding machine for foam injection can have bigger platen for the same locking force.

As the foaming agents decrease the viscosity of the melt, injection can take place with less pressure. Alternatively, the temperature of the melt can be lowered but this measure has an adverse effect on absorption of the foaming agent in the polymer.

Desirable additional features for a moulding machine, specifically meant for structural foam moulding with chemicals, are static mixers and screws with barriers. These measure enhance the homogeneity of the melt and consequently the uniformity of the cellular structure of the moulding.

A shut-off nozzle is mandatory with the injection moulding machine as otherwise the melt may be forced out of the barrel by the gases under pressure.

Gravimetric or volumetric dosing systems for the master batch make the mixture more homogenous, which fact is reflected positively in the product.

The Injection Mould

The moulds for structural foam moulding are essentially identical in design to those for compact injection moulding. As the pressures involved are much lower than in the later case, non-ferrous metals like aluminium and alloys may be used for mould construction. It offers manifold advantages. The higher heat conductivity of aluminium helps reduce the cooling period. The machining time for aluminium is also much less than that for steel and the mould too is much lighter, facilitating easy handling. The cooling can be laid out quite close to the cavity surface to make it more effecient as the mould is not subjected to very high compressive forces.

Though the moulds for foam moulding have ejection and feeding systems like those for conventional injection moulds, the hot runner nozzles must be equipped with positive type needle valves to eliminate leakage of the melt under gas pressure when the mould is in the opened state. Also the sprue bush of the hot runner system needs an effective closure.

The runners and gates are required to be larger than those with compact moulding because the pressure driving the melt is quite low. The moulds must be foreseen with very efficient venting also out of the same consideration. Entrapped air can hinder complete filling of the mould and good reproduction of surface details. Hence adequate escape paths for the enclosed air must be provided at the points where the air would be driven to by the expanding melt.

The surface of a moulding with foam is not as smooth as that of a compact one. A structured surface design can camouflage this visual flaw. The moulds for this process are never polished. If not structured, the cavity surface of the mould is kept rough.

Advantages of Moulding with Chemical Foaming Agents

- Weight reduction due to lower density.
- Cycle time reduction.
- Elimination of sink marks.
- Higher stiffness and bending strength.
- No local stresses near gate.
- Uniform shrinkage, less than that with compact moulding.
- Lower heat conductivity, better thermal insulation.
- Bubbles arrest the propagation of cracks.
- Less chances of degradation due to lower melt temperature.
- Less stresses hence less distortion.
- Possible to mould walls thicker (5–15 mm.) than by compact moulding.
- More dimensional accuracy, narrower tolerances.
- Lower mould cost.
- Lower machine cost.

Disadvantages

- Lower noise absorption.
- Unattractive surface finish, silvery streaks due to ruptured air bubbles.
- Solid residue from decomposition of the foaming agent can cause surface blemishes.

Foaming with Fluids

Though economical in equipment investment, foaming with chemical has some drawbacks. On decomposition, the chemicals release not only gases but also some solids which can deposit on the surface of the moulding and produce blemishes. The level of foaming is also limited. The alternative process of foaming with fluids does not suffer from these limitations.

In the process employing fluids, a pressurised gas, nitrogen or carbon dioxide, is injected into the melt in the metering zone of the plasticising unit of a modified injection moulding machine. The high pressure in the barrel leads to the complete dissolution and uniform dispersion of the gas in the melt. The rotating screw homogenises the melt-gas mixture.

The melt is injected into the mould, filling it partly. The finely distributed gas nucleates and creates microcellular structure. The outer layer of the polymer in contact with mould walls remains compact and smooth. Viscosity of the foamed melt is lower than that of the pure melt with the result that the expanding melt flows further and fills up the mould completely.

As the gas pressure in this process is much higher (150–200 bar) than that generated by the decomposing chemicals in the chemical foam moulding process, articles with wall thickness lower than 5 mm can also be moulded satisfactorily. The higher pressure also leads to higher dimensional stability. The wall thickness, the size of bubbles and the density of the product can be regulated through moulding parameters to some extent.

The Equipment

The injection moulding machine for moulding with pressurised gases differs from the conventional moulding machine primarily in respect of the plasticising unit. The screw is longer, with an L:D ratio of about 24:1 to 28:1. A mixing section forms the end of the screw. The barrel is equipped with one or more gas injection nozzles in the metering zone. The last section of the barrel is equipped with cooling blowers along with the heaters so that the temperature of the melt-gas mixture, which has lower viscosity, can be adjusted down. The nozzle is invariably of the positive shut-off type to prevent leakage of the melt under the gas pressure.

As obvious, the moulding machine is a specific one and cannot be employed for conventional processing.

The gas generation unit does not form a part of the machine though the controls for the gas may be integrated in the machine control panel.

The above description relates to the machinery being manufactured by Engel of Austria. It must, however, be added that many other variations of the equipment have been developed by other machine manufacturers. They all adhere to the basic principle of injecting the fluid into the melt before its entry in the mould.

Alternative Equipment for Fluid Foaming

A way to circumvent the drawback of investment in a special machine is to devise an independent fluid injection module which can be coupled with a conventional injection moulding machine. It has to fullfill the following functions and criteria for successful operation:

1. Load the fluid into the melt at the end of the plasticising phase.
2. Prevent the melt under pressure from getting into the gas injectors and pipelines.
3. Provide adequate surface area for diffusion of the fluid within a short length.
4. Ease of coupling and maintenance.

One of the more successful designs in use has been developed by the Institute of Plastics Processing Aachen (IKV) of Germany. The device (Fig. 11.1) consists of a sintered sleeve housed in a torpedo which forms an intermediate connector between the barrel of the plasticising unit and the machine nozzle. The system forms the path of the melt through the porous sleeve, which is injected with the foaming fluid from outside as well as from inside. The sleeve distributes the fluid all along its area, thus loading the melt with fluid, not at one point as in the previous case, but all along its internal and external surface. A static mixer incorporated in the nozzle homogenises the mixture which remains in single phase till it

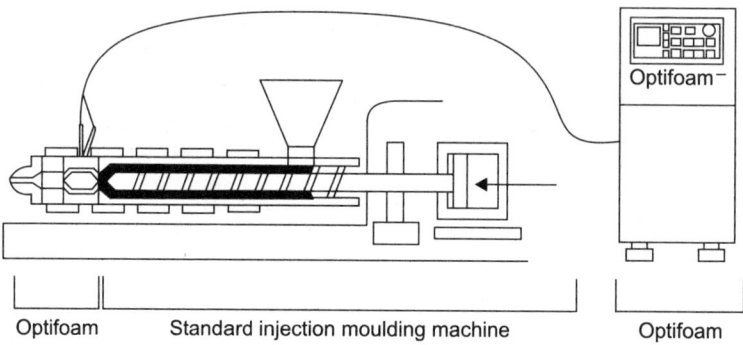

Fig. 11.1

enters the mould. The fall of pressure in the empty mould creates a thermodynamic instability which in turn leads to a two phase mixing. The gas nucleates, builds cells and creates a foamed structure. It also takes over the function of the holding pressure. Figure 11.1 represents the complete unit and figure 11.2 illustrates details of the special attachment.

The system is being marketed by Sulzer of Switzerland.

The Mould

The injection mould for fluid foaming is essentially the same as for the method of chemical foaming. Here too, the pressures involved are much lower than those with the compact moulding process. Light metal alloys with higher heat conductivity than that of the steel for mould plates serve the purpose adequately and also decrease the cycle time, which is further shortened by the absence of a separate holding time. The cooling network can be placed closer to the mould cavity walls.

Advantages of Foam Moulding with Fluids

Basically, the moulding process with fluids offers the same advantages as that with chemicals. However, as only gases are

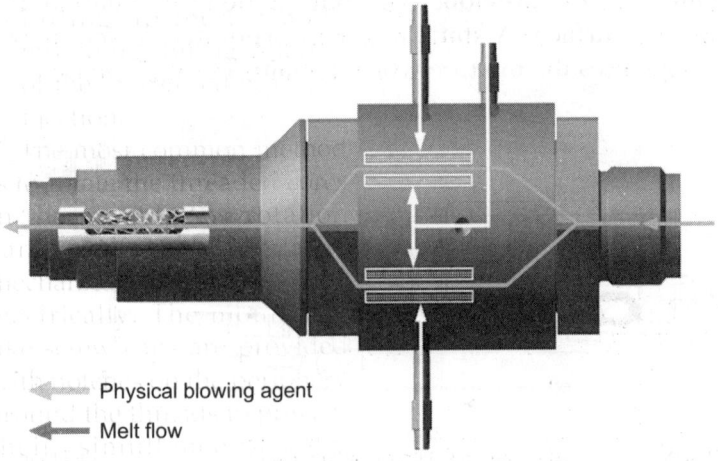

Physical blowing agent

Melt flow

Fig. 11.2

injected in the melt, there is no solid residue left which could give rise to surface blemishes. The proportion of gas and its pressure can be regulated to adjust the final composition of the melt and through that, the inner structure of the moulding.

The main disadvantage is the cost of the special equipment which can only be used for foam moulding. The special screw does not permit the use of the machine for conventional moulding. However, with the special attachments like Optifoam, a conventional moulding machine can be employed.

The Product Design for Structural Foaming

Apart from the general rules valid for moulded articles, following guidelines apply to those produced by structural moulding:

- Minimum wall thickness 4 mm. as the skin is 1–1.25 mm. thick; for chemical foaming agents at least 5 mm.
- Thick and thin sections can co-exist without leading to sink marks.
- All bends and corners well rounded.

Fields of Application for Foamed Mouldings

Packaging and furniture industry, medical appliances, housings of electrical and electronic gadgets, large containers for material handling, pallets and thick walled large articles for diverse fields where the surface finish is of secondary importance, sports goods, toys, etc.

12

Gas-assisted and Water-assisted Injection Moulding

- The process
- The equipment
- The product Design
- Fields of application
- External gas moulding
- The materials
- The mould
- Advantages and disadvantages
- Trouble shooting
- Water injection technique

THE PROCESS

The compact injection moulding, viz. the traditional process, has some limitations; it cannot produce thick walled mouldings without sink marks and voids. Furthermore, there is a size limit to the moulding of thin walled large components. These shortcomings can be overcome by the process termed as Gas Injection Technique (GIT). The thick walled component is rendered hollow by injection of gas in the interior, creating a hollow core so that the product retains its outside dimensions without having accumulation of material which would lead to sink marks and voids.

The process is practiced in a few variations as described below.

The Short Shot Method

In its basic version, the GIT resembles the sandwich moulding (Chapter 9). The mould cavity is filled partially (50–95%) with the plasticised polymer (Fig. 12.1) and then the second component—in this case, a neutral gas is injected in the fluid core of the melt, either through the second channel of a special machine nozzle as shown in Fig. 12.2 or through a separate injection device, which drives it further, pushing it outwards to the cavity walls and filling the mould as it happens with sandwich moulding. The gas pressure, which acts as a constant

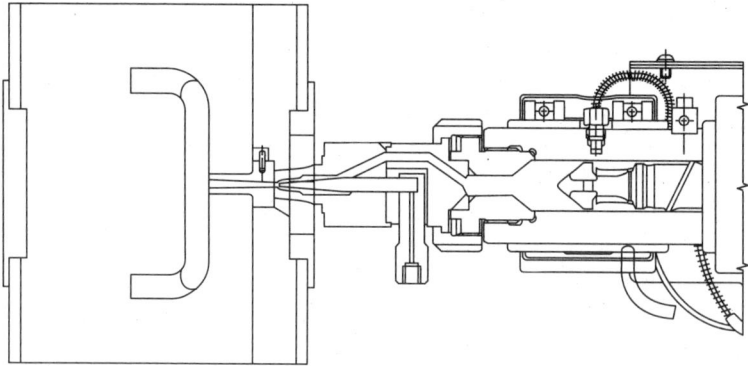

Fig. 12.1

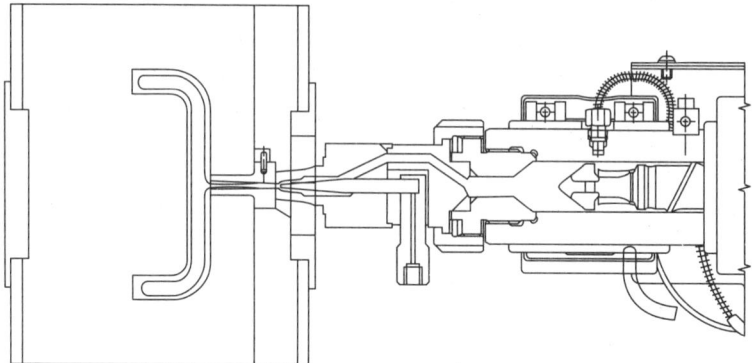

Fig. 12.2

holding pressure, is maintained till the moulding has cooled to a rigid, undeformable article. The gas is evacuated before opening of the mould, either to the atmosphere or back to the storage. The various stages of the process are:

1. Partial filling of the cavity with melt.
2. Injection of compressed gas after the pressure in the mould has reduced below the gas pressure.
3. Holding phase with the gas.
4. Evacuation of the gas.
5. Ejection of the moulding.

To ensure exact (partial) filling, the screw moves till the end of the cylinder, leaving no material cushion.

The pause between injection of melt and the introduction of gas leaves a mark on the surface of the moulding. However, injection of the melt and the gas cannot be undertaken simultaneously as it causes turbulence.

The gas pressure may be in the range of 25 to 300 bar.

The process in the described form is also termed as the Standard GIT.

As obvious, the sprue gate after removal would leave a hole on the moulding. It can, however, be covered by a second injection of the melt as practiced in sandwich moulding (see also 9.1).

Gas Injection Technique with Plastics Expulsion

As pointed out before, the gas injection technique of partial filling may cause surface blemishes, the so-called hesitation marks or splays, which are generated because of the time gap between the end of melt injection and the start of gas injection when the melt is pushed up from the core to the surface. A variation of the basic process to eliminate this defect comprises filling the cavity completely with the melt (Fig. 12.3) and then

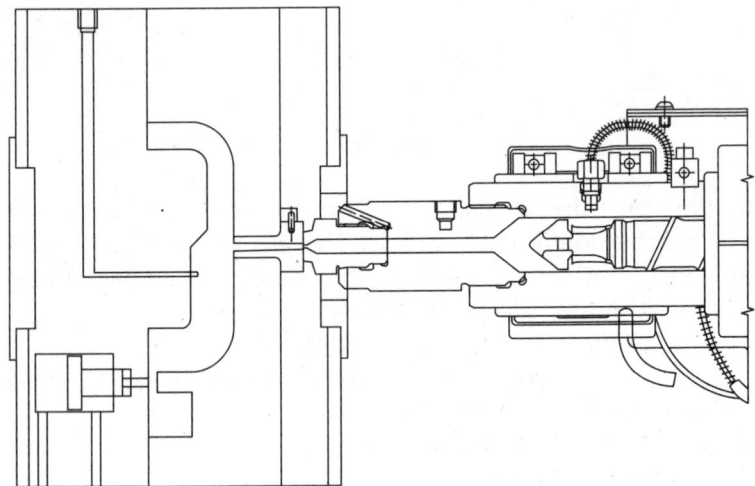

Fig. 12.3

injecting the gas into the melt core. The melt, displaced by the gas, is led to an auxiliary cavity, also called an overspill cavity, the inlet path to which is opened before gas injection (Fig. 12.4). The two diagrams portray this version of GIT in successive steps.

The main advantage of the process is elimination of the surface defects. Moreover, as the cavity is filled completely, holding pressure can be applied and thin sections brought to solidification before injection of the gas. With this, the undesireable finger effect can also be hindered.

The process is accompanied by some drawbacks and disadvantages. The moulding machine has to plasticise and inject as much material as with conventional injection moulding. The expelled material may be counted as a loss. The walls of the final product are generally thicker than with the partial moulding GIT process. Consequently, the cooling time is also longer. The mould, too, is more complicated with additional cavities and the shut-off arrangement.

In order to salvage the expelled material, the overspill cavities may be in the form of some article of use either with the main product or elsewhere. However, very little control is possible over its quality as all parameters have to be set for the main cavity.

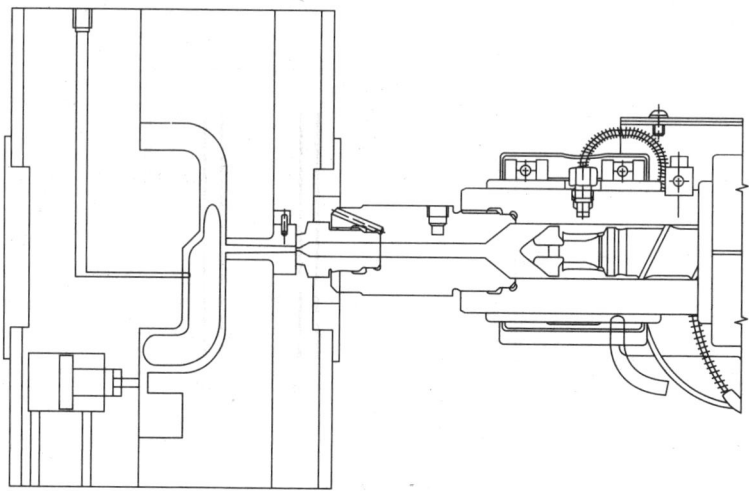

Fig. 12.4

Gas Injection Technique with Melt Push Back

This method offers an alternative to transferring the melt to overspill cavities, thus saving material besides delivering a surface finish free of blemishes. The   process is similar to the forgoing one with the difference that the injected gas pushes the melt back into the plasticising unit while creating a hollow path (Fig. 12.5). There are, however, some stringent limitations to the process and the product which prevent its universal application.

- The gas has to be injected at the last point of the melt path.
- The product cannot have side branches.
- It is practicable only with single cavity moulds. There can be only one injection point.
- The return path to the plasticing unit, viz. the sprue and runners, must remain molten.
- The gas may penetrate into the plasticising unit.
- The cooling time is longer.
- The material is susceptible to thermal damage.

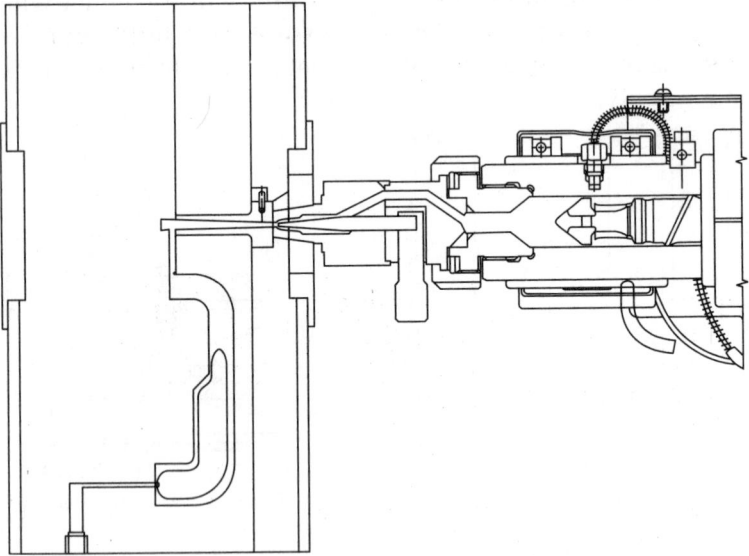

Fig. 12.5

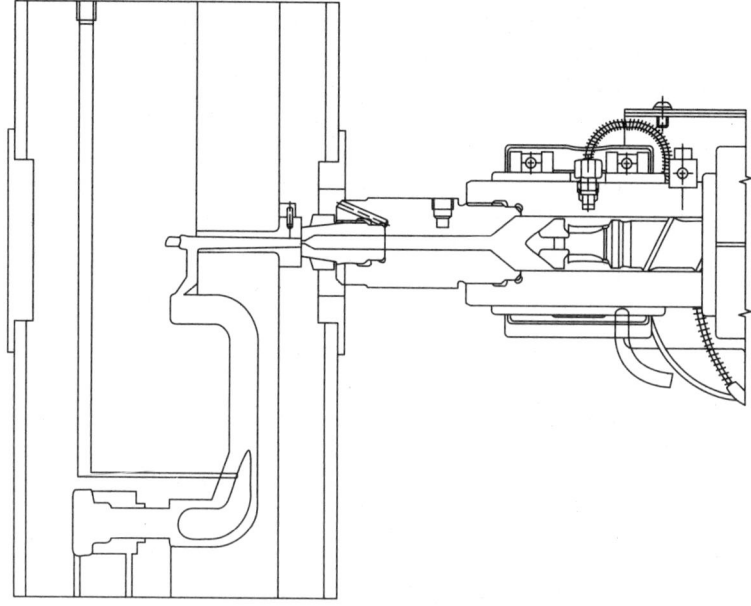

Fig. 12.6

Gas Injection Technique with Retractable Core

Yet another variation of the basic GIT process consists of filling the mould cavity having a retractable core, with the plastics melt completely when the core is in the front position. In other words, the volume of the cavity is reduced at the time of filling. The core is retracted simultaneously as the gas is injected. The space vacated by the core is filled by the melt being pushed out by the gas.

Because of complete filling without interruption, the surface finish of the article is uniform and there are no marks of the melt coming out on surface after injection of the gas.

This variation of GIT has limited application. All articles do not offer possibility of incorporation of cores. The retracting cores leave their mark on the surface of the article. The mould too is more expensive. Figure 12.6, with the core in pulled-back position, illustrates and explains the process.

The setting of machine parameters, especially the start of the gas injection, its pressure profile, the pressure, temperature and viscosity of the melt at the time of gas injection as well as the mould temperature exercise a decisive effect on the surface quality, length of the channel hollowed by the gas, wall thickness of the moulding and its uniformity. Fillers influence rheology of the melt and consequently the wall thickness. They raise the thermal conductivity of the melt and accelerate the heat dissipation. The faster cooling results in thicker frozen layer. The geometry of the article too influences the resultant wall thickness.

In view of the myriad variables, the optimal parameter setting is still a matter of trial and error.

Process Variations

There are several variations to the basic process, developed and patented by different companies and bearing different names. Some of them are:

- GIT   Gas Injection Technology
- GAIM   Gas Assisted Injection Moulding
- GAM   Gas Assisted Moulding
- GAIN   Gas Assisted Injection moulding
- GID   Gas Innen Druck (English: Gas Internal Pressure)
- GIP   Gas Injection Process
- CINPRES Controlled Internal Pressure moulding
- EGM   External Gas Moulding
- WIT   Water Injection Technique

All processes named above, except the last two, are based on the same principle, viz. creating hollows inside the moulding by injection of a gas. Developed by different manufacturers, they differ in details, equipment and conditions of licensing.

THE MATERIALS
Plastics

It is possible to mould practically all thermoplastics and also some thermosets (they must have a fountain flow) with the gas injection technique. Thermally sensitive materials like PVC are prone to degradation, especially with blow-out and push-back versions of the process. Filled and unfilled thermoplastics, amorphous as well as semi-crystalline, have been successfully moulded. Grades with lower melt flow indices are better suited for the process as the melt has more strength and the risk of the gas blowing through is lower.

The most common materials processed commercially with GIT are PP, HIPS, ABS-PC blends, PA 6 and PA 66.

The Gas

Compressed air, though the cheapest alternative, has proved hazardous because of the oxygen component. It reacts with certain plastics and leads to complications.

Nitrogen, although more expensive, is absolutely neutral, inert and safe as it does not react with any of the plastics. It is in abundant supply, forming major part of the air. It can be extracted with suitable equipment in house or purchased in cylinders or tanks. Its disposal poses no problems as it can either be let out directly in the atmosphere or re-routed to storage for reuse.

A major drawback of nitrogen is its very low specific heat or the cooling capacity. Nitrogen can also be chilled before injection to expedite cooling. It needs special cooling equipment and additional relief valve for venting of excessive pressure generated when the liquid gas turns into vapour.

THE EQUIPMENT
The Injection Moulding Machine

In GIT too, a normal injection moulding machine forms the key unit. The accuracy of the product is directly dependent on the accuracy of the shot volume, particularly in the basic process. A closed loop machine yields more consistent results. As the injection takes place without a material cushion and the tip of the screw touches the end of the barrel bore, it is vital that the non-return valve functions impeccably. Any leakage will have an adverse effect on consistency of production.

Injection of the melt and subsequently that of the compressed gas can be carried out in various manners. The method, which does not call for a modification in the basic moulding machine, involves injection of the melt through a shut-off nozzle, whereby the filling takes place through a runner and a side gate. The nozzle is closed and the gas is injected by means of a separate, independent gas injection nozzle placed at a convenient point in the cavity or in the runner. The disadvantage in the former case is that the hole left by the gas injector cannot be overmoulded. The location for gas injection is chosen so that it is either invisible in normal position of the component in use or it does not influence the function adversely.

The Special Nozzle

Gas injection moulding can also be carried out by equipping the moulding machine with a special nozzle having separate flow paths for the melt and the gas (Fig. 12.7). As described in section 12.1, the injection takes place through a sprue gate. After injection of the

Fig. 12.7

predetermined volume of the melt in the mould cavity, the path of plastics in the nozzle is blocked and that for the gas is opened when the pressure prevalent in the cavity has come down below that of the gas. The pressurised gas pushes the fluid material from the core forward and upwards, filling the cavity completely. Thereafter, the gas acts as the medium for the holding pressure till the moulding is rigid enough to be ejected without distortion.

The machine controls have to be modified for operation of the nozzle and injection of the gas.

Gas Injection Equipment

Nitrogen can be obtained in cylinders or tanks. It has to be compressed to the required level, usually to 300 bar and in rare cases up to 400 bar. It needs pressure regulation during injection and holding phases. In other words, a gas injection unit consists of:

a. Compressor
b. Pressure regulator
c. Timers and controls

The controls of the gas equipment must be synchronised with those of the injection moulding machine. Special machines have an interface for that but it is possible to incorporate it in the normal moulding machines also. Machines, especially built for GIT, have all relevant controls for the gas integrated in their control systems. The switch-over from melt injection to gas injection can take place either time bound or distance related. It can also be connected with the machine hydraulic pressure or the melt pressure in the mould cavity over a pressure transducer built therein.

GIT specialised injection moulding machines have also the provision for storage of data of a number of mould settings.

In case of regular consumption on a large scale, it is more economical to generate nitrogen in house with the help of special equipment, which would extract the gas out of air, compress it and regulate it. Depending upon the size, the equipment can act as central supplier to a number of machines. The regulation takes place individually through a pressure regulation module with electrical controls, which can also be mounted on the mould.

Gas Injectors

Unless the gas is injected into the melt through the special machine nozzle as illustrated in Fig. 12.1, provision must be made in the mould to introduce it either in the runner or directly in the mould cavity. It is executed by incorporating an injector in the mould. Its simplest version is a slender hollow tube with a core pin which leaves an annular gap of 0.02–0.04, which is wide enough for the gas to pass but too narrow for the melt to get in. The core pin moves up and down with flow of the gas and performs a cleaning operation with each stroke.

The narrow-gap injector may get chocked if used with "melt expulsion GIT" because of the high melt pressure involved for filling the cavity completely.

The injector is housed in the cavity plate in direction of ejection.

A variation of the above design is a hollow tube closed with a sintered plug. The porosity of the plug enables passage of the gas. The plugs get clogged after some time and must be changed or cleaned. The design of the porous injector facilitates a ready exchange of the plug.

A more sophisticated version of injector consists of a hollow outer tube with a movable central mushroom shaped valve, which is kept closed during injection of the melt by a spring (Fig. 12.8). When pressurised gas is injected through it, the valve opens and the gas drives the melt from the core in the unfilled parts of the mould cavity, creating thereby a hollow in the moulding. The injector also serves as a conduit for gas evacuation. To this end, the lower end of the

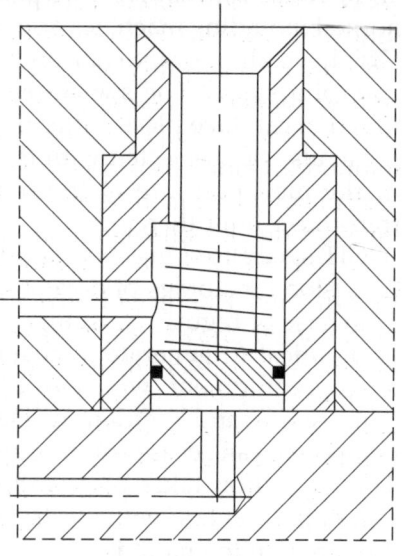

Fig. 12.8

valve is shaped like a piston of a pneumatic cylinder. Pressurised gas, introduced through a channel below the piston pushes up the valve, forming an exit path for the gas from the moulding.

A dual purpose injector nozzle, developed by Cinpres, has a spring loaded mushroom valve which keeps the injector tube closed till it is pushed open by the incoming gas. For evacuation of the gas, the injector nozzle assembly, consisting of the valve, tube and its holder at the base, is pulled back by a small hydraulic cylinder fitted at the back of the nozzle, thereby exposing the exhaust path covered by the nozzle base in the front position. Figure 12.9 illustrates the mode of operation in principle.

The configuration of the product may not permit location of the injector in direction of the mould opening. Injectors placed

at an angle to the axis of
opening must be retracted
before ejection stroke can
begin. The retraction can
be effected mechanically,
pneumatically, hydrauli-
cally or electrically.

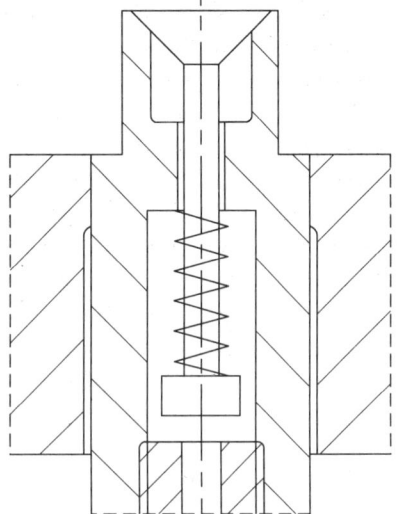

*Injectors offer several advan-
tages over special dual purpose
machine nozzles such as:*

- Small size
- Flexibility in choice of
  optimal location for gas
  injection
- Ease of assembly and
  replacement
- Incorporation of more
  than one injector if
  dictated by product configuration

Fig. 12.9

- Possibility of gas injection in multi cavity moulds
- Saving in cost compared to special nozzles
  The tubular injectors leave a small hole in the product.

THE MOULD DESIGN

Gas assisted injection moulding can be rightly regarded as an
extension of the standard injection moulding process.
Consequently the moulds for GIT are essentially similar to those
for the parent process with suitable modifications.

The mould for the process variation with partial filling,
carried out with dual purpose machine nozzle, is a single cavity
one. It is identical to a standard injection mould in all respects.
The core and cavity plates can, however, be fabricated out of
non-ferrous alloys as the pressures involved are low. The higher
heat conductivity of these alloys contributes also to the cooling
efficiency.

Partial injection GIT moulds with separate gas injection can
be of multicavity type. The feeding may be through
conventional runner network or also by means of a hot runner

system with shut off nozzles. The crucial requirement is the accuracy and uniformity of the shot in all cavities. It is extremely important to fill the cavity homogeneously so that the gas bubble can be guided as designed in order to achieve uniform wall thickness. Fill the cavity from bottom to top against the force of gravity to achieve a swell flow. The machine nozzle should be of positive closing type.

As far as possible, the gas injector should be positioned in direction of ejection. It is generally housed in the cavity plate on the moving half of the mould. Most standard injectors have a threaded base for assembly. Their size permits their accommodation between cooling lines. The rule for their location is that the gas should drive the melt towards the unfilled extremities of the cavity. The injector should be placed in the vicinity of injection so that the gas bubble follows the melt flow and expands in one direction only. More than one injectors should be employed sparingly and only in situations when the gas channels remain independent.

If the configuration of the article makes it unavoidable to place the injector across the axis of ejection, arrangement has to be made to pull it back before the mould opens or before ejection starts. It may be achieved mechanically, pneumatically or hydraulically as the case may be.

The timings of start of the gas injection, its pressure, volume, duration and exhaust are set on and regulated by the gas control unit.

Injection moulds for GIT are cooled in the conventional manner. Any thin sections of the article, which should not be blown, should have more intensive cooling so that it is solidified before the gas is injected. The measure helps prevent the finger effect.

The relatively low pressure of the gas, which is used to fill the cavity completely, makes it imperative to provide efficient venting in the mould so that filling can proceed evenly and cavity details are reproduced faithfully. Entrapment of air can lead to usual defects.

Injection moulds for melt pushback technique are essentially the single cavity type as driving back the melt from more cavities into a single source will cause mutual hindrance and blockage. The location of the gas injectors must drive the melt in the direction of the sprue. In all other respects, the mould is similar to a cold runner mould for standard injection moulding.

Injection moulds for the melt expulsion process are of two types, viz. with overspill cavities of fixed size and alternatively of variable size. They are linked to the main cavity through a gate-like small channel which may be an open or shut-off type. Its size and shape influence filling as well as gas propagation.

From aesthetic angle, it is desirable to have the "gate" as small as possible so that the unavoidable mark on the main product does not mar the appearance. Too small a channel may, however, give rise to jetting. It may also retard the flow of gas and thus give rise to thicker walls. Too large a gate tends to let the melt out in the overspill cavity before injection of the gas and freeze and close prematurely.

It is recommended to keep the length of the gate as short as possible, say about 1 mm. The depth and breadth may be 2–3 mm. The edges should be rounded off.

The shut-off gate employs a sluice operated by a pneumatic or hydraulic cylinder controlled by the machine. Its operation is linked with the injection of the gas.

The overspill cavity with fixed volume does not add to the complexity of the mould but its size must be pre-calculated accurately. The variable volume overspill cavity, on the other hand, provides freedom in setting of parameters to achieve the desired wall thickness but necessitates moving devices such as a hydraulic cylinder to operate the slide varying the volume of the cavity. The cylinder has to be operated and controlled in relation with gas injection. It bestows a measure of flexibility but also adds to the cost of the mould as well as to the complicacy in setting.

Demoulding takes place predominantly with ejector pins of adequate size. These also serve as self cleaning venting devices. Ejectors are also placed under the auxiliary cavities, which should be provided with adequate draft for safe ejection.

THE PRODUCT DESIGN

The gas injection technique is primarily suitable for two types of moulded articles.

- Thick walled, long and slim unidirectional articles like hollow rods, closed tubes, handle bars, table legs, pedals, etc. (Fig. 12.10).
- Thin articles with large area, foreseen with thick walled gas channels (Fig. 12.11).

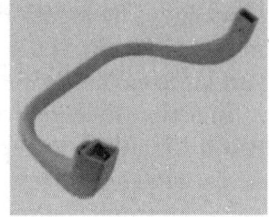

Fig. 12.10

Longitudinal Articles

The guiding principle for design of articles to be moulded by GID is the shape favourable to the fluid flow. The most natural path that a fluid under pressure, be it a gas or a liquid, would carve out for itself in a plastics melt is round. A fluid can push

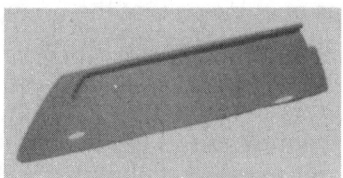

Fig. 12.11

out the melt which comes in its path in a directed manner but side branches cannot be controlled effectively. Some guidelines for an effective product design are given below.

- As the gas forms preferably a circular channel, the outer contour of the product should also be as close to a circle as possible so that the wall thickness is nearly uniform all over. Material accumulation leads to same defects with GIT as with standard injection moulding (Fig. 12.12).
- If a circular cross section is not feasible, the outer corners should be rounded off generously to reduce material accumulation (Fig. 12.13).
- If the product profile consists of multiple sections, the meeting points of the sections or the

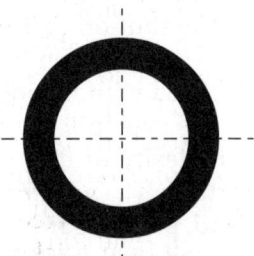

Fig. 12.12

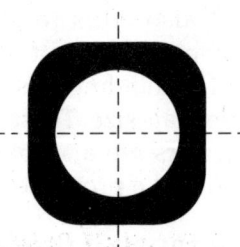

Fig. 12.13

transition zones should be foreseen with as large radii as possible. Sharp corners yield thin walls (Fig. 12.14).

- Bends represent critical areas. Sharp bends result in uneven wall thickness whereby the wall on the inner side may become too thin and that on the outer side too thick. Select the greatest possible bend radius (Fig. 12.15).

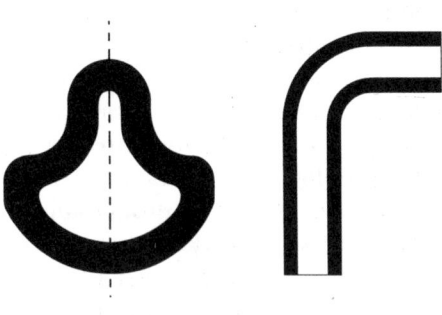

Fig. 12.14   **Fig. 12.15**

- Maximum width of a moulding can be 3–5 times its height. The length should not exceed five times the height.
- The ratio of the article wall thickness and the gas channel breadth should be 1:2 to 1:3

Large Flat Articles

Thin walled articles with large areas, when injection moulded in standard manner, call for very high injection pressure and large machines with high clamping force. The resultant mouldings contain stresses which cause distortion. Multiple gating, which is often unavoidable for filling, brings weld lines and weak spots. Such products, suitably modified, can be easily moulded in GIT, without the defects cited above.

The redesign of the articles involves incorporation of ribs which discharge a dual function. They stiffen and strengthen the flat area and serve also as channels for conveying the gas. The pressurised gas pushes the melt out and thus fills the initially incompletely moulded article.

The advantages are obvious. As the thin-walled article need not be moulded completely, the injection takes place at lower pressures. Consequently, a smaller machine with lower clamping tonnage may be employed. Simultaneously, lower injection pressure results in lower stresses and less distortion. The gas pressure, which remains same at all points unlike the holding pressure, ensures an overall uniform shrinkage. Following are some guidelines and rules for design of GID compatible large, flat articles:

- Avoid material accumulations; provide for uniform wall thickness for the entire moulded part.
- Thick-walled large areas are unfavourable as the penetration of gas progresses uncontrolled and undefined. Finger effect, that is, breaking away of gas in several small channels from the main stream, is quite likely.
- Reduce wall thickness of a large article and provide ribs which strengthen and transport gas. The article wall thickness should be uniform and not exceed 3 mm. as far as possible. The root of the rib is designed as a structure which acts as gas channel. Likewise the corner, where two walls meet, can be fashioned as a gas channel. Figure 12.16 illustrates some examples.
- The thickness and height of the ribs are correlated to the thickness of the article.
  a. The thickness of the rib (t) should lie between 0.5 to one times the thickness of the wall (T) it stands upon. Adequate taper for ejection must be foreseen.
  b. The height of the rib (h) above the gas channel may vary from 5 to 10 times its thickness.
  c. The gas channel at the root of the rib may be 2 to 3 times as high (H) as the wall thickness ($H = 2 - 3 \times t$). The width (W) of the channel should equal its height. The outer corners should be well rounded.
- The junction of the channel and the product wall should be provided with small radii to eliminate the notch effect. A big radius increases the wall thickness and with that, the risk of the gas penetrating sideways in the product wall.

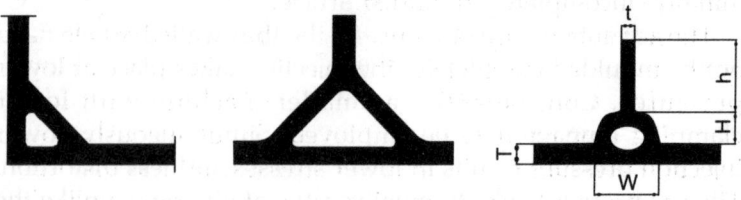

Fig. 12.16

- The ribs should be so placed that gas and the melt are driven towards unfilled extremities of the cavity (Fig. 12.17).
- Channels closer to the gate would take the melt farther than those away from it, resulting in uneven filling. The flow can be equalised by varying the size of the channels.
- Labyrinthine ribs and those laid out in zigzag or grid pattern should be avoided as far as possible. Gas bubbles proceeding from two directions may cause choking, air traps, burn marks and sink marks at locations where gas could not reach.

Functional Elements

Theoretically, it is possible to integrate functional details like fastening lugs, brackets, hinges, pins and holes, snap fit elements, etc. in the articles for GIT moulding but these are likely to cause problems during cooling and shrinkage. These form zones of material accumulation but unlike in standard injection moulding, these cannot be taken care of with compensating holding pressure. Sink marks, deformations and dimensional inaccuracies become unavoidable without additional measures such as individual gas injection.

Holes across the flow of the gas can be made as in conventional moulding by means of sliders but the cross pins divide the gas bubble unpredictably. It may lead to critical and unacceptable wall thickness variations.

In conclusion, it may be underlined that many of the processing faults encountered in GIT are the outcome of

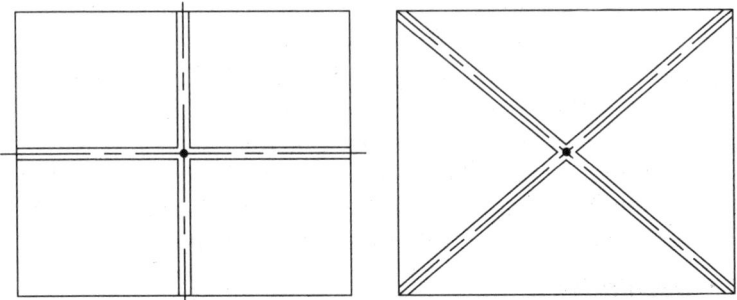

Fig. 12.17

inadequate product design. It is often very difficult to remove or subdue the defects by changing processing parameters.

ADVANTAGES OF THE GAS ASSISTED INJECTION MOULDING

- Elimination of sink marks and voids in thick mouldings
- Reduction of weight, material saving in the range of 20–30%. More rigidity with less weight due to hollow structure.
- Large components can be moulded with fewer gates and fewer weld lines.
- Cheaper hot runner aggregate because of fewer nozzles.
- Faster moulding cycle because of less material to cool.
- Greater stiffness with less material by dint of hollow ribs.
- The gas is a million time less viscous than the melt and does not lose pressure through flow, hence uniform holding pressure through the gas at all points, which is lower and remains effective till the end independent of gate closure, consequently uniform shrinkage, lower stresses and less distortion.
- 90 % of the gas can be retrieved and reused.
- Closer tolerances because of uniform holding pressure and uniform shrinkage.
- Lower clamping force, smaller injection moulding machine with less energy consumption.
- Cheaper moulds made of non-ferrous alloys because of lower forces.
- Less material to plasticise hence lower energy consumption
- Less cooling time because of less material and better heat conducting moulds.
- Long flow paths possible.
- More flexibility of product design; possibility of combining thick and thin sections. Elimination of assembly of separately moulded solid and hollow components.
- Possibility of including slim inserts as the pressure is low.
- Possibility of creating parts with hollow channels for fluids etc. (A shower head is moulded in this way)

Disadvantages

- Additional cost for the gas unit and the gas.
- Changes in the controls of the moulding machine.

- Not suitable for very complex products with features mouldable at high pressures.
- Hole at the point of gas injection.
- Wall thickness is not uniform.
- Splays on surface in partial filling variation.
- Possibility of difference in surface finish over the gas channel.
- Surface blush over the gas channels.
- Loss of material in the melt expulsion process.
- Higher mould cost in blow-out and retractable core processes.

APPLICATIONS

- Automotive: bumpers, grilles, door panels, spoiler, gas pedal, mirror housing, instrument panels, glove box, motorcycle shield, etc.
- Aerospace: interior and structural panels.
- Consumer electronics: TV cabinets, back covers, computer housing.
- Appliance cabinets, refrigerator and washing machine panels, top covers.
- Garden furniture.
- Housings for electrical components.
- Office equipment panels and housings.
- Sanitary equipment like shower head, toilet seat, bath handles.

FAULTS AND REMEDIES

Switch-over Marks

There are lines on the surface of the moulding at extreme ends of the melt charge. These constitute a visual but not a structural fault.

In GIT with partial filling, there is always a time lag between the end of the melt injection and the start of gas introduction. The surface of the melt in contact with the mould walls and at the ends of charge get cooled and forms a rigid layer. As the gas pushes out the molten material out of the core to the surface, the frozen ends of the the melt front are pushed up against the mould wall but cannot assume the surface texture. This causes marks at the joint of the stationary charge and the pushed up material. It is more or less a hallmark of the process.

The time lag, the root cause of the fault, is unavoidable as the gas can be injected only when the melt pressure has come down to a level lower than the gas pressure. The marks may be reduced, shifted to some less conspicuous part of the article or hidden by:

- Changing the shot volume.
- Increasing the melt temperature and reducing the injection pressure.
- Increasing the mould temperature.
- Increasing the gas pressure.
- Enlarging the runners and gates.
- Changing over to a better flowing material.
- Choosing material with lighter colour.
- Removing or covering the marks in after-operations.

Finger Effect

The fault occurs predominantly with large, thin walled articles foreseen with gas channels. The gas leaves the predetermined path and carves out irregular side channels. The surface finish over the fingers is distinctly different.

The gas follows the path of least resistance. If it encounters resistance while blowing the partly filled cavity, it will make ingress into areas which offer less resistance. The product design may have built-in pitfalls which facilitate incursions of the gas in undesired regions.

The finger effect can be avoided by:

- Dimensioning the gas channels as recommended in the product design guide
- Not exceeding the prescribed wall thickness limitations for the thinner section
- Providing very small radii at the joints of ribs with the flat surface
- Extending the time gap between melt injection and gas injection
- Decreasing the gas pressure or setting it in a decreasing profile

- Reducing the gas holding time
- Using a thermoplastics with lower shrinkage (amorphous thermoplastics are less prone to finger effect).

The Slip-stick Effect

The defect manifests itself as a scaly surface of the product.

As the process of melt displacement by the pressurised gas proceeds and the amount of melt being pushed decreases, the speed of the gas can achieve such proportions that the frozen material already sticking to the mould wall is pulled loose, dragged along and deposited further.

The remedy lies in:
- Reducing the gas pressure
- Creating a pressure profile during the holding phase
- Increasing the melt charge
- Raising the mould temperature

The Diesel Effect

Burn marks indicating charred material.

The defect and its cause are the same as in conventional injection moulding. The entrapped air is compressed by the advancing melt front and gets overheated resulting in burning of the material in contact. The phenomenon may occur with standard GIT with partial filling.

The defect can be eliminated by:
- Adequate venting through shallow slots at the location of entrapment
- Strategically placed ejectors or sintered metal inserts
- Decreasing the gas pressure and thus slowing the speed of the front, so that air gets time to escape
- Selecting alternative location of injection to prevent formation of two flow fronts which are coming in the way of air escape

Glass Fibre Streaks

White streaks showing glass fibres on the outer surface. The defect is more prominent in materials of dark colours.
- It can be circumvented by using higher gas pressure.
- It can be rendered less conspicuous by using a lighter colour.

Sink Marks

Sink marks are unforeseen depressions on the surface of the moulding.

The cause of sink marks is same as with conventional injection moulding. Material accumulations cause sink marks during cooling if the material deficit created by shrinkage is not compensated through input of additional material.

A major cause of material accumulation is the faulty product design. Some solid functional details like assembly elements, fixtures, hubs and studs may need wall thickness which produces sink marks unless compensated. The gas pressure is generally too low to perform this task. Material accumulation may also take place at the extreme end of the gas channels, at the bends and besides cores in the path of the gas. The fault can be countered by:

• Reducing the shot volume
• Providing separate gas injection for thick details
• Decreasing the gas pressure
• Increasing the gas holding time
• Designing the bends with larger radii and a larger angle
• Increasing the size of the overspill cavity to reduce rest material at the end
• Increasing the cooling period
• Cooling the sink mark prone areas intensively to transfer the void inwards.

Distension of the Moulding

The article blows up after ejection.

As a rule, the gas is evacuated and the product has the normal atmospheric pressure also inside before the mould is opened but the gas is a poor cooling medium and the poor heat conductivity of plastics retards heat dissipation to the cooled mould walls so that the interior walls may still be in semi-molten state. The escaping gas tears along minute material particles which choke the narrow annular gap of the injector. Runners and gates too may no more be open when gas is evacuated. The trapped gas causes the article to blow out or even burst.

Following measures help eliminate the problem:
- Longer cooling period
- Colder mould
- Larger runners and gates
- Lower gas pressure
- Retarded exhaust rate
- Frequent cleaning of injectors or gas channel of the machine nozzle

Gas Channel Imprint

The difference in surface finish over the path of the gas in the article.

The channel cored out by gas does not have a uniform wall thickness. The difference in thickness is accompanied by different shrinkage and also difference in reflection of light by the surface of the moulding. It is particularly distracting when the gas does not follow the intended path, e.g. wanders out in flat sections.

The most effective remedy is revision of the article design. Some improvement can, however, be achieved by:
- Varying the gas pressure to reduce wall thickness difference
- Reducing the shot volume to decrease material accumulation
- Increasing the gas holding period

EXTERNAL GAS MOULDING

Unlike other gas assisted processes, the EGM does not use gas to create hollows in the moulding. The process is more like injection compression moulding whereby the gas, injected below the article, forms a micro-thin layer and compresses the melt against one face of the mould to counter emergence of sink marks on that side.

Large thin articles with ribs can be moulded successfully with standard GIT but the marks left by the gas channels on the opposite surface, usually against the ribs, may not be acceptable in many cases. EGM provides a solution here by pressing the entire surface of the moulding against the mould face forming the visible side of the product till the end of solidification. Unlike the normal holding pressure which is higher near the gate and diminishes with the distance and time, the external gas pressure is uniform and constant. The process eliminates the sink marks and surface finish differences on the intended face entirely or to a great extent depending upon the product design and the plastics.

It must be understood that sink marks are not eliminated; they are transferred to the less critical face such as the inner side of a housing or a cover.

The process of external gas moulding takes place in four distinct steps (Fig. 12.18). After complete filling of the cavity (a), compressed gas is introduced in the mould, from the

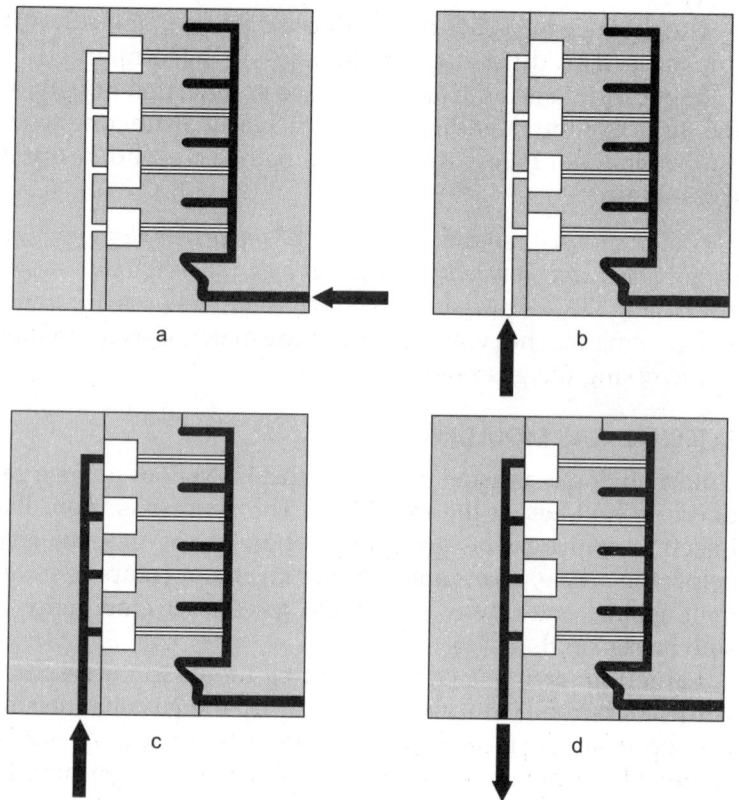

Fig. 12.18a to d

opposite side of the visible face (b) and forms a micro-thin layer under a high pressure which may range between 50 and 300 bar. The injection mould has to be designed to permit access of the gas over the entire face and hold it in place till expulsion. The parting face of the mould must be provided with a seal. The gas takes over the role of holding pressure with the difference that it acts uniformly over the complete area till the end (c). The gas equipment is the same as for other GIT processes. The gas is evacuated before the mould is opened (d).

Apart from eliminating sink marks on the surface, the process manifests other improvements like reduced moulded-in stresses, more uniform shrinkage, narrower tolerances, less distortion and better flatness—an attribute highly important in case of mating components. The process also proves highly effective for moulding large flat articles like flat panels for office equipment, tops for domestic appliances, furniture components like table tops, outdoor furniture parts, wash tubs and automobile panels, etc. There may be a reduction in the cycle time and also in the weight as no additional material is pushed in during the holding phase.

EGM offers more freedom in product design as thicker functional details can be integrated without the risk of visual defects.

THE WATER INJECTION TECHNOLOGY

The Process

The basic difference between the GIT and WIT is that the later employs water instead of a gas as the medium to core out the moulding. The process was developed by IKV, the Institute for Plastics Processing, University of Aachen, Germany and taken over by Battenfeld, the moulding machine manufacturer in the same country. BASF, the plastics raw material producer, has tested various materials for their suitability for WIT and has developed some special grades of materials for processing with this technology.

The process is almost identical to the gas aided injection moulding. Pressurised water is introduced through injectors situated in the mould cavity, filled partially or fully with plastics melt. All variations of GIT except the one with dual purpose machine nozzle are valid for WIT also. The pressure of water is about 300 bar. Before opening the mould for ejection, water is evacuated either through gravity or expelled by blowing air or gas through another nozzle built in the cavity for this purpose.

For evacuation of water by gravity, it is imperative that water injection takes place from the lowest point in the mould. It necessitates placement of the article in the mould accordingly. The melt should also flow against the gravity in the mould cavity to ensure good surface finish.

WIT has proved especially advantageous for two types of articles. These are:

- Pipelines for fluids (Fig. 12.19)
- Structural components stiffened by means of hollow channels

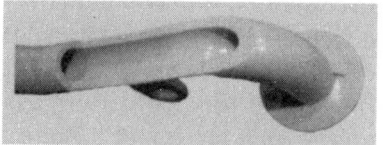

Fig. 12.19

Materials

WIT has been almost exclusively employed for production of technical components. Consequently, engineering thermoplastics have been tested, modified and employed commercially for mass production. Predominant polymers are PA 6 as well as PA 66, with or without glass fibres. PBT too has found application in some cases.

ADVANTAGES AND DISADVANTAGES

Advantages

Water as the agent to core out the moulding offers several advantages over the gas.

- Water is cheaper than the gas.
- Its heat conductivity is forty times higher than that of nitrogen so that it exercises a faster, direct cooling from inside. The cycle time is considerably shorter, almost a quarter of the GIT process.

- It has higher viscosity, almost equalling that of the polymer melt with the result that the wall thickness distribution is more uniform. It yields thinner walls as compared to gas.
- Water injection yields a much smoother internal surface than the gas. The fibres are not cored out.
- Water facilitates moulding of combination of thick and thin sections.
- Water is incompressible. It can core out very long sections as its front acts as a solid ram against the fluid melt, pushing it farther and also cooling it in the process.
- The cooling time can be shortened, especially in case of thick-walled mouldings, by maintaining an inflow of fresh water in the product by means of two separate water nozzles.
- Water injection moulding eliminates the danger of foaming.
- The water injection moulding offers a viable alternative for blow moulding in certain cases, especially for fluid-carrying components.

Disadvantages

- Unlike gas, water cannot be injected through a common machine nozzle. It needs a separate water injection nozzle.
- Water must be injected rapidly to achieve an even wall thickness distribution and to yield a smooth surface inside. It necessitates large channels for its flow. Consequently, the water nozzles are much larger than their counterparts for gas injection and need more space in the mould.
- Water nozzles must have a shut off arrangement. The function is executed hydraulically, which contributes to their complexity, size and cost (Fig. 12.20). The moulding machine, too, must possess the additional hydraulic circuits with relevant controls.
- Water leakage causes greater nuisance than the gas leakage.
- Water must be removed from the moulding

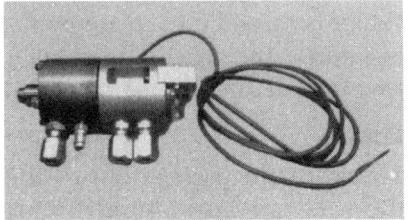

Fig. 12.20

before ejection. It may necessitate additional arrangement to inject compressed air in the article to expel water. That calls for an additional injection opening.

* The water injection technology requires a process specific water compression unit with all relevant controls to bring up a pressure of 300 bar.
* WIT has proved economical mostly in case of thick walled long components.

Applications

* Seats, back and arm rests for cars and buses
* Car mirror holder
* Car roof railing
* Wiper arm
* Pipe fittings, ducts
* Furniture components like chair legs
* Sports goods, tennis rackets
* Conveyor rollers
* Handles
* Machine components like chain saw handles

Faults and Remedies

Some faults which are typical to the WIT process are:

Surface Waviness and Whirls Close to the Water Injection Point

The defect is caused by too high a water pressure and can be eliminated by starting with a lower pressure and increasing it to the required level quickly.

Partial Coring by Steam

Water is turned into steam by the heat of the melt which cores out partly the article. It occurs when the quantity of water is too small. It can be remedied by increasing the volume of water.

Pores

These are tiny depressions on the inner surface and below it. If the water pressure builds up very slowly, a very thin layer gets frozen which gets punctured susequently by the high pressure.

The fault can be eliminated by adjusting the rate of pressure build up.

Voids

Voids are the hollows or cavities in the wall of the article, invisible from both sides.

These are caused by the cold walls of the mould and cold water under lower pressure. Thick layers get frozen both inside and outside. As the water pressure cannot compensate the shrinking material between the frozen layers, voids are formed in the walls.

The solution of the problem lies in increasing the water volume during injection and maintaining high water pressure during the cooling phase.

Conclusion

The process is undergoing continuous development. A combination of gas injection with that of pressurised water has extended the palette of plastics which can be processed by WIT. In this variation, a bubble of gas is introduced before injection of water in the melt. This acts as a cushion between the hot melt and cold water, preventing sudden freezing and yielding much smoother surface besides reducing turbulence in water flow where the article section changes.

13

Powder Injection Moulding

- The process
- The equipment
- The product design
- Applications
- The materials
- The mould
- Advantages

THE PROCESS

Injection moulding of powders, although a form of the conventional moulding of the thermoplastics, differs from the normal process in some vital details. The term "Powder Injection Moulding" stands for moulding of articles out of powdered solid materials rendered mouldable with the help of a thermoplastics carrier. If the powder is that of a metal, the process is termed as MIM. Likewise, in case of ceramic powders, it is called CIM. In its entirety, the process encompasses:

- Preparation of the feedstock
- Injection moulding of "green compacts"
- Debinding to "brown compacts"
- Sintering

Preparation of the Feedstock

Fine ceramic or metal powder is mixed with a binder, which is invariably an organic compound, mostly a thermoplastics. The function of the binder is, as its name suggests, binding the powder particles together. The primary condition that a binder must fulfil is that it should wet the powder particles, encompass them completely and act as a holder without reacting with them. It must also be completely removable without effecting the shape of the article formed out of the powder. Other desirable characteristics are good mould removal properties, dimensional stability and adequate green strength after moulding. It should be thermally stable but must decompose easily.

The proportion of the binder is kept to a minimum possible, generally between 30 and 40 per cent in volume, depending upon the type of filler and the binder. The aim is to keep shrinkage, resulting from the binder removal, as low as possible. This ensures closer dimensional tolerances in the finished product. Waxes, lubricants and wetting agents as well as pigments in the form of powdered metallic oxides may also be added. The mixture is homogenised and palletised on an extruder. The resultant granulate forms the feedstock for injection moulding.

All metals and ceramics, which can be powdered and sintered, are suitable for powder injection moulding. The powder grains should preferably be spherical in form and as small and uniform in size as possible. Coarse grains cause higher friction in the injection unit and reduce its service life.

Readymade feedstocks with different metals and ceramics are now available commercially.

Injection Moulding

The process of injection moulding of powder feedstock is essentially the same as for the thermoplastics granulate. The peculiarity of the raw material, viz. the very high contents of non-melting abrasive fillers and very low amount of the melting thermoplastics binder has influence, not only on the moulding parameters but also on the design, material and construction of the moulding equipment and the mould.

The process of plasticising, viz. feeding, compression, melting and metering is exactly the same as in case of thermoplastics. The low contents of thermoplastics in the feedstock, however, cause less expansion of the melt and permit very little compression. The screw speed is also kept low as the low amount of the plastics needs less shear. The circumferential velocity should not exceed 15 meters per minute. It should be lower for ceramic powder moulding.

The feedstocks with powders have a significantly higher viscosity than the binder thermoplastics. Consequently, the injection and holding pressures would also be on the higher side. The fillers make the melt highly heat conductive. It is therefore necessary to fill the cavity as fast as possible. Jetting must be avoided, preferably through proper gate positioning and if that is not possible, through lower filling rate in the beginning. The temperature profile set on the barrel corresponds to that suitable for the binder plastics. The nozzle temperature should, however, reduce the melt viscosity The holding pressure is kept up till the gate freezes. The melt cools comparatively fast. A thumb rule for the cooling time of metal filled moulding is that it is as many seconds as the square of the thickest section of the article in millimeters.

The mouldings are referred to as "green compact" (Fig. 13.1) and are prone to distortion during ejection. They should be handled with robots and stacked in suitable trays providing adequate support and protection against deformation. The trays must be suitable for use in the subsequent operations of debinding and sintering.

Fig. 13.1

Debinding

Debinding is the operation of removing the thermoplastics which has acted as the carrier and the binder of the powder that was to be given a foreseen shape by injection moulding. It has fulfilled its function and now it must be eliminated in such a way that the moulding retains its form and can be subjected to the final operation of sintering.

Different binders are employed for different powders. The method of removal deper.ds upon the type of the thermoplastics used in the particular case. The binders are extracted by way of catalysis, dissolution or thermal decomposition. Binders consisting of polyolefin-wax mixtures are removed by melting. The process has to be carried out at a very low rate to prevent distortion caused by expansion in volume of the binder on melting. Some binders can be removed by disolving them in organic solvents. Polyalcohols and polyvinylalcohols, which are soluble in water, are most suitable for this method.

Polyacetal, the most common binder, is eliminated by thermal decomposition, supported by gaseous nitric acid. The process is carried out in a debinding oven. The green compacts to be debinded are put on oven grids in the trays these were placed in after demoulding. Nitrogen creates an inert atmosphere in the oven, which is heated and maintained at 110°C. Nitric acid is introduced after the green compacts have been warmed up. The decomposed binder releases formaldehyde in gaseous form which is extracted and burnt out.

The debinding rate at 110°C varies between 1–2 mm of the article thickness per hour. The process can be carried out in a long, tunnel type oven at appropriate speed so that the articles emerging at the other end of the tunnel (now they are called brown compacts in case of metals and white compacts with ceramics), are ready for the next operation. A brown compact is illustrated in Fig. 13.2.

Fig. 13.2

About 80–90% of the binder is eliminated by debinding. The rest is required to keep the filler grains together for safe transport to the sintering oven.

Sintering

In order to firmly bind the particles of the brown compact and to turn it into a pore-free component, it is sintered by application of heat. The process is the same as that for the press-sintered articles as far as the furnace atmosphere and the temperature-time curve are concerned. The temperature may range from 1300°C to 2000°C depending upon the material being sintered but it is raised gradually in the beginning to drive away the rest of the binder. The cooling, too, is effected in small steps of 5–10°C per minute.

Injection moulded sintered parts can approach the theoretical material density up to 99.9%. They

Fig. 13.3

would have all properties of the parent metal when properly debinded and sintered. A sintered article in its final shape is shown in Fig. 13.3.

The three steps of production from moulding to sintering are usually carried out in a row as a continuous process as shown in the schematic diagram 13.4.

The production process may be regarded as well-balanced when the final product has a uniform structure and appearance, is free of pores, has a smooth surface and possesses all properties of the parent material.

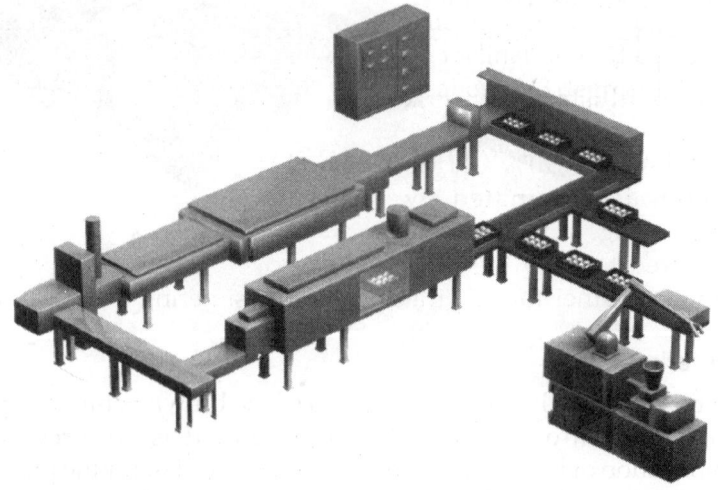

Fig. 13.4

The Shrinkage: The moulded components shrink in two steps:

The mouldings shrink after ejection and cooling. The shrinkage occurs in the binder thermoplastics but not to the extent as when it were unfilled. A fact worth taking into account is the expansion of the heated mould. At 130°C, the mould dimensions are larger by 0.2%. Consequently, the moulding too is bigger by that amount. The net moulding shrinkage is therefore less.

The operation of sintering leads to further contraction as the filler particles come closer and fill up the gaps left by the extracted binder. With suitable binder, the brown compacts shrink isotropically, i.e. evenly in all directions during sintering. The shrinkage after sintering may be 20% or even higher. This has a significant effect on achievable tolerances.

Other factors which influence shrinkage are:
- The binder contents
- Powder characteristics
- Mixing process and quality of mixing
- Moulding parameters
- Distortion due to gravity
- Sliding properties on the sintering underlay
- Uniformity of the sintering process

Dimensional Accuracy

Typical tolerances for powder injection mouldings are:

Nominal limit in mm.	Tolerance ± mm.
<3	0.05
3–6	0.06
6–15	0.075
15–30	0.15
30–60	0.25
>60	± 0.5% of nominal value

The Materials

Among metals, copper, hard metals, stainless steel, strontium ferrite and platinum are some of the materials processed successfully by the powder moulding technology.

The achievable particle size depends upon the hardness of the material. It may vary between 1 and 20 μm. A spherical form of the particles causes less friction and wear of the processing equipment.

The palette of ceramics processed by CIM includes aluminium oxide, porcelain, zirconium oxide, silicone carbide as well as silicone nitride. The particle size varies between 15 and 45 μm.

Polyolefins and polyacetals are the most commonly used binders.

Additives are waxes, lubricants and wetting agents as well as pigments in the form of powdered metallic oxides.

The material is recyclable. A 50:50 blend of reground and virgin material can be safely used for the same moulding.

THE EQUIPMENT

The Injection Moulding Machine

The injection moulding machine for PIM differs from its conventional counterpart primarily in geometry of the screw of the plasticising unit. The low contents of the thermoplastics in the feedstock permit very little compression of the melt. Consequently, the screw can have only a small compression ratio. It is usually in the range of 1:1.16 to 1:1.2.

The metal as well as ceramic powders are highly abrasive. Unless suitably hardened, the components of the plasticing unit will wear out very soon. It is particularly true for the non-return valve. Any back leakage caused by a worn-out valve would result in inconsistent shot volume and loss of pressure during the holding phase, which may prove detrimental to the dimensional accuracy of the moulding. The valve components should preferably be made of sintered carbide.

A closed loop machine would yield more uniform quality.

A robot for taking out the ejected mouldings and positioning them for the subsequent operation of debinding invariably forms a part of the moulding unit.

THE MOULD

The injection moulds for powder injection moulding are subject to the same guidelines of design as their counterparts for thermoplastics. However, the altered nature of the melt, such as its increased viscosity and the high proportion of the fillers and their abrasiveness make certain modifications unavoidable. The feeding system has to be more generously dimensioned because of the sluggishness of the melt. A good polish facilitates the flow and demoulding. All parts coming in contact with the abrasive fluid must be through-hardened. If possible, the gate section, which is likely to get worn-out fast, should be easily replaceable. Sintered carbide inserts for gating have a longer service life. In case of polyacetal based feedstock, it is safer to employ hardenable stainless steel for crucial mould components.

The very high content of the filler leads to a reduction in shrinkage, necessitating generous taper for a trouble free demoulding. A minimum taper of 1 degree is a must though a larger one up to 3 degrees would be preferable. It must be borne in mind that the mouldings do not possess much rigidity because of the low amount of the binder and are liable to distort during ejection. It is not only the generous draft angle but also adequate area of ejectors which helps avoid distortion and damage during demoulding.

The feeding system merits special attention because the very high content of fillers raises the viscosity considerably. Following measures keep the pressure drop within reasonable limits:

- Runners as short as necessary.
- Runner cross-section as large as possible.
- Runner cross-section preferably circular.
- No sharp bends in feeders. All junctions generously rounded.
- Gate at the thickest section but no free flow of the melt stream.
- Gate size bigger than that for the unfilled resin.

All known gate types, including the submarine gate, can be employed. Hot runner systems, too, can be used. In fact, it is easier to insulate the hot runner unit against heat losses here as the entire mould is maintained at a high temperature. However, provision must be made for abrasive nature of the melt. Complex needle valve nozzles are unnecessary since the melt does not tend to drool or string during demoulding. The gate tears off cleanly.

The flow pattern of the melt in the mould is identical to that of unfilled thermoplastics. A bad mould design and wrong moulding parameters can give rise to moulding faults like weld lines, warpage and imperfect reproduction of the surface and the finish. These defects cannot be remedied in the subsequent operation of sintering. They must be eliminated at the stage of moulding only.

The moulds are maintained at an elevated temperature during the moulding process by circulating a hot fluid through a network of holes or a labyrinth, incorporated in the mould for this purpose. The layout of the network must ensure uniform temperature at all points. The maximum deviation should not exceed ± 2°C. A larger variation manifests itself as differences in surface finish, strength and dimensions. The high mould surface temperature is necessary as the filled melt loses heat rapidly due to the high heat conductivity of the fillers. The mould temperature may have to be set as high as 140°C for feedstocks with polyacetal as binder. The seals should withstand the temperature.

The injection moulds for the powder injection moulding can have all usual features such as three plate feeding system, moving, sliding, revolving or collapsible cores as well as unscrewing devices and other mechanisms, common in moulds for thermoplastics.

A vital feature of the mould is the venting arrangement. Entrapped air acts as a brake to the even flow of the melt. Apart from the well known diesel effect with burnt material, the trapped air may lead to inadequate reproduction of the surface details.

The resistance offered to the advancing front by the enclosed air being progressively compressed may slow it down and result in incomplete filling due to fast cooling of the melt. In most cases, venting slots on the parting face, 0.01–0.02 mm. deep at the end of the flow path, prove effective. However, depending upon the geometry of the article, air may get trapped at points away from the parting line. In such cases, auxiliary ejectors, positioned at the crucial spots, offer escape route to the air.

Though seemingly a minor point, the prevention of loss of heat from the mould is no less important for maintaining a constant temperature in the mould. Insulation sheets between the mould clamping plates and the machine platen help save heat energy and reduce temperature fluctuations.

THE PRODUCT DESIGN

Although the final product would consist of metal or ceramics only, the process of its formation, viz. injection moulding, subjects its design to the rules valid for moulded plastics articles. Some applicable guidelines are:

- To counter distortion, the walls of the product should have a uniform thickness.
- Avoid abrupt wall thickness changes.
- Employ ribs instead of thickening the walls.
- Core out thick sections. Maximum wall thickness < 20 mm
- Avoid sharp corners. Minimum corner radius - 0.3mm.
- Provide a generous draft angle for safe demoulding.
- As far as possible, one face of the product should be designed flat to seat the green moulding.

The powder injection moulding process can also be employed for moulding products in more than one material as well as in conjunction with gas injection. The prime condition is similarity of shrinkage and sintering temperatures.

The MIM products can be subjected to usual post operations like sand blasting, grinding, polishing, galvanising, plating, soldering, welding, machining and hardening.

ADVANTAGES

The process offers following advantages:

- Production of ready to use components. Post-operations in rare cases.
- Possibility of surface treatment and decoration after sintering by physical, chemical and mechanical methods.
- No restrictions on design intricacies.

- High surface quality.
- Dimensional stability and repeatability.
- Possibility of full automation.
- Recyclability of material. No wastage.

APPLICATIONS

Metals

Watch bodies, small gears, components for hand and machine tools, weapons, medical and dental tools, computers and printers etc.

Ceramics

Thread guides, sealing washers, components for pumps, valves, motors, turbine wheels, heater cores, bearing sleeves, spray jet nozzles, etc.

14

Injection Moulding of Thermosetting Plastics

- The materials
- The equipment
- The product design
- Advantages and disadvantages
- The process
- The mould design
- Trouble shooting

THE MATERIALS

Thermosets were the first fully synthetic plastics. Baekland brought out Phenol Formaldehyde under the name of Bakelite way back in 1909. Since then, they have been processed by compression moulding and transfer moulding processes which are essentially semi-automatic, labour intensive and consequently expensive. Despite their outstanding properties, thermosets have been losing ground to the more rationally processible thermoplastics. Development of the injection moulding process for thermosets in 1965 has given a new lease of life to them. It has enabled the duroplasts a spectacular comeback. Their inherent properties have secured them a firm position in certain branches of industrial applications. The thermosets have:

- High static and dynamic strength
- High electrical and thermal insulation
- High thermal and chemical resistance
- No softening after curing
- No halogens, sulphur and heavy metals
- Very high wear resistance
- The capability of replacing non-ferrous metals

The Form of the Moulding Compound

Injection moulding depends upon gravity feeding of the machine with material. The compound should, therefore, be free flowing. In other words, it cannot have rough type of fillers. The suitable fillers are wood flour or refined sawdust, cotton flock, short glass fibres not longer than 0,5 mm., crystalline calcium silicate, etc. These fillers reduce the material cost, lend heat resistance, improve electrical properties and enhance strength. It also has certain additives, which retard cross-linking at lower temperatures to prevent hardening during the dwell time in the injection unit but accelerate curing at higher temperatures.

The material should also have as little dust or powder as possible. It may hamper free flow and cause irregularities in the charge.

Injection moulding machines for moulding of dough and bulk compounds are equipped with arrangements for forced feeding.

Summing up, the thermosetting compound for injection moulding should have the following attributes:

- Uniform granular form of a size between 0,5–1 mm.; free of dust and powder
- No coarse fillers
- Easy gliding in the hopper for uniform charging
- Should glide easily through the barrel and over the screw
- Slow hardening around 100°C, the temperature of flow in the barrel
- Fast curing at temperatures above 150°C.

Storage of Thermosetting Raw Materials

As the heat favours curing or crosslinking, the packed and sealed bags of duroplasts must be stored in cool rooms, less prone to temperature and humidity fluctuations. Out of this consideration, there should be no heating or water pipes passing through the stores. The bags should not be in touch with the cold floor either as the moisture is likely to shift towards the colder part of the packaging. The ideal conditions for storage are:

- The ambient temperature about 20°C and the relative humidity around 50–60%.

- The shelf life of duroplasts is always limited. It is advisable to store only the quantity needed for a short time and keep on replenishing it.
- Before processing, the material from the stores should be conditioned by keeping it in the production rooms for sufficiently long time, at least a day or more.

THE PROCESS

Injection moulding process for thermosets is similar to that for thermoplastics in many ways. In a plasticising unit consisting of a barrel and a revolving and reciprocating screw, with controlled heating, the material is brought to a state where it can flow under pressure without undergoing any appreciable curing, that is, crosslinking. It is injected by the screw acting as a plunger into a heated mould which has one or more cavities of the shape of the required product. After complete filling, the pressure is reduced but kept on till the gate solidifies. The screw of the plasticising unit can start the charging for the next shot at this stage by rotating and receding. The moulding can be ejected when it has cured to an extent that it would not distort during demoulding. The internal chemical reaction process continues outside the mould till completion. The individual phases of the process are described below.

The material in the form of granules is fed into a hopper at one end of the plasticising unit consisting of a hollow steel barrel and a closely fitting screw. The screw transports it further towards the opposite end. The material passes through an intermediate stage of sintering mainly from the heat of friction and from the heating jackets around the barrel of the plasticising unit, if the former does not suffice, till it assumes the form of a viscous fluid. Should the heat generated by the rotating screw be in excess of that required, the heating jackets perform the function of coolers. The increasing pressure exerted by the plasticised material expels the air and gases towards the hopper.

As the fluid material is injected into the mould through a linear stroke of the screw, the pressure needed is that to overcome the resistance offered to the flow by the feeding path and the mould walls (the thicker the article, the lower the resistance) and the viscosity of the material. It is to be noted that the fillers have a far reaching influence on the flow properties. As the melt proceeds further, it receives heat from friction against the mould walls, from the heated mould and from the exothermic reaction it is undergoing. The reaction gets faster as more and more heat is generated. It results in reduction of viscosity and that of the injection pressure. The pressure rises again when the mould is filled up and the compaction starts. Another factor responsible for increase in pressure is the decreasing viscosity as a result of the progressive curing. The material at peak temperature expands and may generate flash if the locking force of the machine is inadequate. Trapped air and gases liberated at high temperature, if not evacuated through efficient venting, may further push up the internal pressure and contribute flash.

As the curing progresses, the material undergoes contraction. The pressure at this stage, viz. the holding pressure, can be reduced. It is generally about 60% of the injection pressure. The viscosity of the material also declines with curing and at one stage, the holding pressure becomes ineffective.

The total time needed for hardening is governed by the maximum wall thickness.

The Machine Setting

Following is a typical setting of parameters for most thermosets.
Barrel temperatures:
Feeding zone 60°C, middle zone 60–70°C, Nozzle 90°C

Screw speed	70–100 r.p.m.
Injection pressure	1400 bar
Holding pressure	600–1000 bar
Back pressure	150–200 bar

The Moulding Cycle

The time taken for injection is relatively short. The friction during injection enhances the flowability. The holding phase too is not much longer. The major part of the cycle is formed by the curing time. It may be as much as 60–80% of the total cycle time. It can be reduced if the compound can be subjected to higher temperatures in the barrel as well as in the mould without reducing its flowability or expediting curing.

The sequences of a moulding cycle are:
- Mould closes
- Injection unit advances
- Injection phase
- Holding phase
- Curing
- Mould opens
- Ejection
  (Charging takes place after the holding phase during curing.)
  Figure 14.1 illustrates various steps of a moulding cycle in relation with one another.

The Heating

Various sources of heat to plasticise and cure the material are:
- Heating jackets around the barrel
- Work done by the screw
- Frictional heat through flow in the runner
- Exothermic condensation reaction
- Heaters in the mould

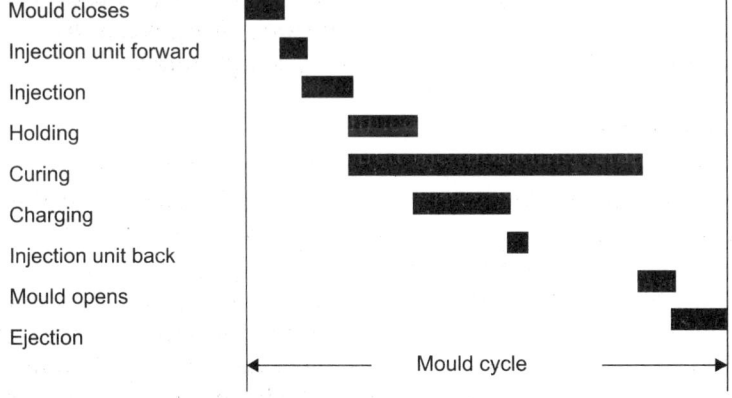

Mould closes	
Injection unit forward	
Injection	
Holding	
Curing	
Charging	
Injection unit back	
Mould opens	
Ejection	

Mould cycle

Fig. 14.1

The phase in the cylinder may be regarded as "preheating". It is a sensitive operation as the material at too low a temperature resists injection and too high a material temperature may initiate the chemical reaction right in the barrel, leading to hardening and blockage. The right temperature of the melt for injection is the one when the fluid has almost the lowest viscosity.

The vital difference from the moulding process for thermoplastics is that the barrel of the plasticising unit for thermosets is heated only to a level where the material is plasticised without undergoing a crosslinking reaction. Secondly, the quantity of the material plasticised corresponds to one shot only. The screw acts mainly as a transporter and mixer. The mould is heated to a much higher level which initiates and facilitates the crosslinking action which in turn hardens the moulding. This, too, is a significant difference from the moulding of thermoplastics. These are made flowable by input of heat and hardened by its extraction. Thermosetting plastics are also heated to make them flowable but it is the chemical reaction and not the cooling which brings them to solidification.

THE EQUIPMENT

The injection moulding machine for thermosets is similar to, and in many respects identical with, that for the thermoplastics. It also has a plasticising unit, a mould closing unit and an ejection unit, performing functions suggested by the nomenclature. The difference, if any, lies in details.

The Plasticising Unit

The plasticising unit (Fig. 14.2) consists of a hollow steel barrel placed horizontally and provided with 2–3 water/oil heating jackets, and in rare cases with electrical band heaters and cooling fans, around its circumference. The heating jacket closer to the feeding zone has lower temperature than the next one. There is usually a separate jacket for the last section adjoining the nozzle as it is here that the friction caused by injection may raise the temperature significantly. The hot mould too transfers

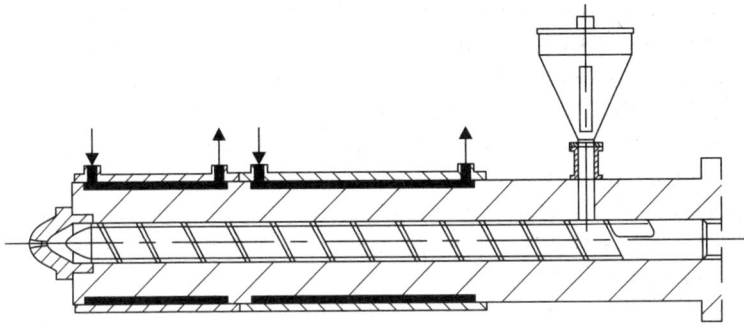

Fig. 14.2

heat to the nozzle, particularly during the dosing operation. The circulating fluid extracts the surplus heat.

On the barrel end forming the start, a material hopper sits on an opening cut in the barrel. It is a conical container made of stainless steel plate. As the filled thermosetting granules do not glide down very easily, the slope of the cone is not more than 40 degrees. The inside of the hopper, too, is well polished to facilitate gliding. The hopper is foreseen with a transparent window to display the level of contents. The other end of the barrel forming the exit is fitted with an open nozzle having the exit hole much smaller than the bore of the barrel. The barrel contains a screw, closely fitting in it and capable of rotation, retraction and forward movement. When it rotates, the raw material falling on it from the hopper is transported forward towards the nozzle. The depth of the flutes and the pitch of the screw are generally constant all along the length and the material experiences no compression during its transport from the feed end to the exit. In some cases, the depth may decrease a little after some distance from the feeding point causing a slight compression of the raw material to expel air. In other words, the screw may have a compression ratio varying between 1:1 and 1:1,1 for different materials. The minimal variation in depth also facilitates removal of solidified material for cleaning. A screw with compression will have three sections, viz. feed, compression and metering zones as with screws for moulding of thermoplastics. The material should not have a long dwell time in the barrel. The screw, therefore, is much shorter than that in plasticising unit for thermoplastics. The screw length varies between 10 to 15 times its diameter. The depth of the flutes is generally one tenth of the diameter. The surface of the screw is highly polished whereas the bore of the barrel is kept rough. This measure helps transport of the material. The difference in temperatures of the barrel and the screw also serves the same purpose. Screws of larger diameters are provided with an internal heating arrangement.

As most thermosetting compounds contain highly abrasive fillers like glass fibres, minerals, wood flour, wood fibres, clay, carbon black, fabrics, etc., the screw and the barrel are subjected to considerable wear and tear. The steels for these machine parts and particularly their heat treatment should take this into account.

It may be underlined that the main function of the plasticising unit is to convey the material to the mould with as little curing as possible. This is why, the length of the unit is less, the compression is very low and the heating is with water as the temperature of the compound in the barrel is kept below 90°C. Out of same considerations, non-return valves which may give rise to friction and overheating, are seldom employed in the moulding machines for thermosets. The backflow of the resin is checked partly by the special design of the screw/barrel head (Fig. 14.3) and by blocking the reverse rotation of the screw. An open type nozzle, screwed onto the end of the barrel, forms the end of the plasticising unit.

Short interruptions in production cycle lead to hardening of the material close to the nozzle in the barrel and longer interruptions to the hardening of the whole charge in the barrel. The nozzle and the screw should, therefore, be easily dismantleable.

· The size of the charge is slightly more than the shot weight. The additional material serves as cushion needed for transmitting the holding pressure. However, with some thermosets like epoxides, which harden readily, no cushion is permitted so that the screw tip, which is formed like a cone as shown, almost touches the corresponding conical cavity of the nozzle at the end of the injection stroke.

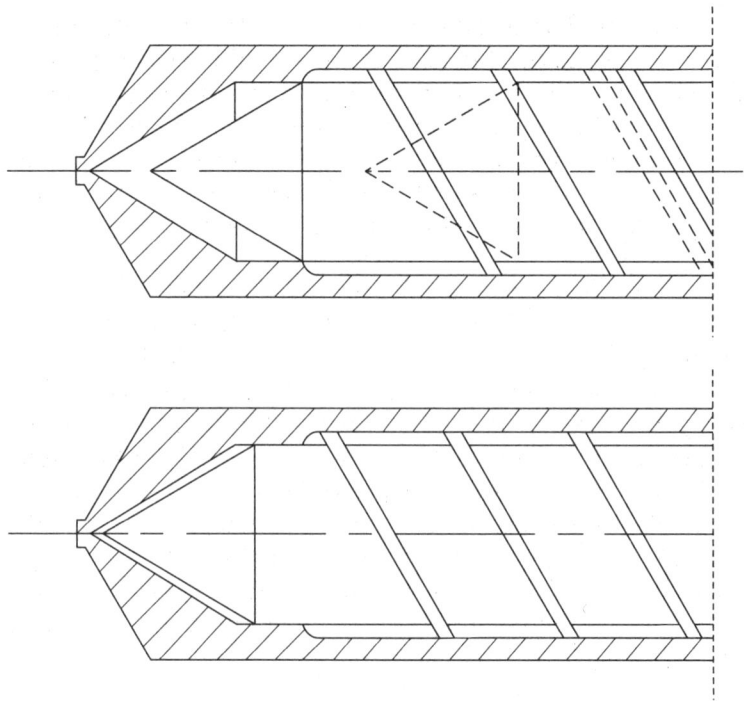

Fig. 14.3

The Mould-closing

The mould-closing unit of a thermoset injection moulding machine is, in principle, a replica of its counterpart in an injection moulding machine for thermoplastics. It consists of a stationary platen, placed vertically in line with and close to the plasticising unit. It bears an accurately formed circular opening in the middle to provide exact location to the mould and access to the nozzle of the injection unit. The platen, which is fixed on the machine bed, has a number of tapped holes for clamping of the mould. A parallel platen placed on the same axis and capable of a linear movement towards and away from the fixed platen, forms the other part of the mould clamping unit. A set of tie bars, usually four in number, joins the platen and guides the moving platen. The moving platen too, has a number of tapped holes for mould clamping and also an accurate round hole in the middle for centring the other mould half. This opening also enables the machine ejector to move the ejector plates of the mould. A third platen behind the moving one, fixed to the bed, supports the tie bars and also serves to hold one end of the toggle lever or the hydraulic system for powering the moving platen. The whole assembly is mounted on a bed which houses oil tank, the motor, the pump, valves and other controls.

With the moulding machines for thermosets, hydraulic cylinders are preferred to the toggle lever mechanisms for imparting movement to the platen and locking it. The hydraulic devices exert force in the middle of the moving platen (Fig. 14.4, right) and are not effected by the increase in the mould height due to heating. As a result, any deflection in the mould due to the closing force and the injection pressure is in the same direction. This is not the case with toggle levers, which depend upon the extension of the tie bars for locking and press on the moving platen away from the central axis. Consequently, the heated mould halves may bend in opposite directions (Fig. 14.4) giving rise to flash.

Ejection

The ejection device is identical with that in injection moulding machines.

Special Features and Ejection

Most modern thermoset moulding machines possess a "breathing" device for the mould closing unit, a pneumatic powered brushing fixture and air blowers for cleaning of the

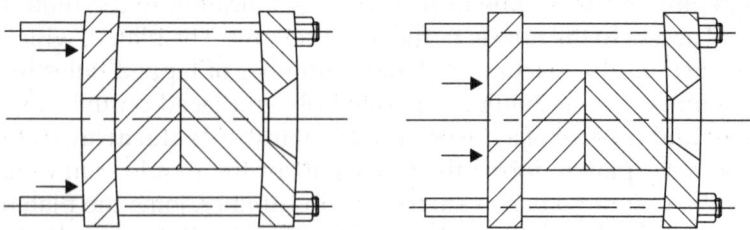

Fig. 14.4

mould. The breathing device provides venting by opening the mould locking unit by 0,01–0,03 mm. right after filling of the mould. The brushes and the air blowers remove the thin fragments of the cured material left by the mouldings after ejection.

INJECTION MOULDS FOR THERMOSETS

The Mould Heating

The moulds for thermosets are not much different from those for the thermoplastics. In fact, they are much simpler as they do not require cooling at all. Instead, they are heated. The heating arrangement for moulds for thermosets usually consists of electrical heating cartridges incorporated in the retaining plates. These should sit snugly to ensure efficient heat transfer. Their main function is to heat up the mould in a short time before the operations can start. Their capacity is estimated on the basis of the weight of steel they have to heat. As a thumb rule, 200–300 watts are provided per kilogramme of the weight. Round moulds can also be heated with band heaters. During the moulding process, a major part of the heat required for cross-linking comes from the friction generated during injection of the material at a high velocity through the runner system. A sophisticated device aiming at maintaining a uniform temperature in all sections of the mould cavity is shown in Fig. 14.5. Strategically placed heat pipes convey heat from the retainer plates, heated by cartridge heaters to the inaccessible parts of the core and cavity. The thermal consistency results in uniform curing, shorter cycle time and less warpage.

The Feeding System

Apart from the fact that the sprue and runners cannot be reground and reused, the feeding system in injection moulds for thermosets is no different than that employed in injection moulds. The shape of the sprue may be identical to that for thermoplastics. The diameter is usually bigger. In order to reduce the wastage, either a special machine nozzle going deep in the mould may be used or better still, the sprue bush may be provided with cooling from outside and/or from inside

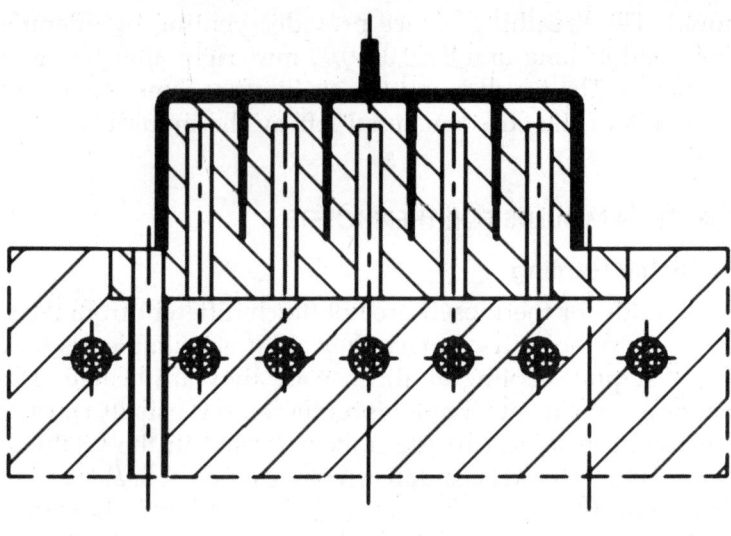

Fig. 14.5

(Fig. 14.6. The cooled sprue remains practically uncured and need not be taken out. The runners, too, may be shaped like those for moulding of thermoplastics. A circular cross-section is ideal but the runners may also be trapezoidal or rectangular. The three-plate system can also be employed if required. The main difference in concept is that here there is no solidified insulating layer protecting the fluid "core". The runners should be smooth and well polished. In general, the runners should not be thinner than the thickest section of the moulding. They should be as short as possible to avoid premature hardening. If a mould does not fill well under normal conditions with an

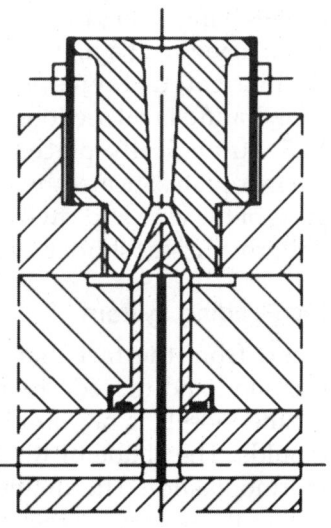

Fig. 14.6

injection pressure of 5 tonnes, then the feed system is too restricted, the feed path too long and/or its cross section too small.

The secondary runners should be smaller than the main one in order to maintain the material under pressure, the proportion being very much a matter of experience as the material does not behave like a true hydraulic fluid. The longer the flow path, the more viscous it becomes.

The entrance to secondary runners must have generous radii to assist the flow to change direction.

The Cold Runner System

Analogous to the hot runner system of injection moulds for thermoplastics, a cold runner system has been developed for thermosets to keep the runner temperature low and thus reduce or totally eliminate the wastage of the material forming runners. The whole network of runners is accommodated in a separate block, housed in the mould but having very little physical contact with it (Fig. 14.7). The block is efficiently cooled to keep the level of cross-linking as low as possible. The part of the mould housing the core and cavity is provided with heating.

Like the hot runner standard units for thermoplastics moulds, ready-made cold runner composite systems for

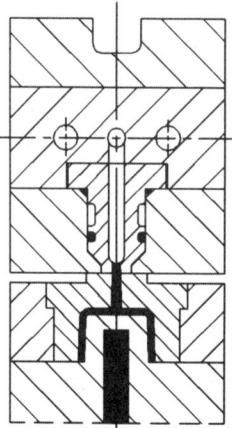
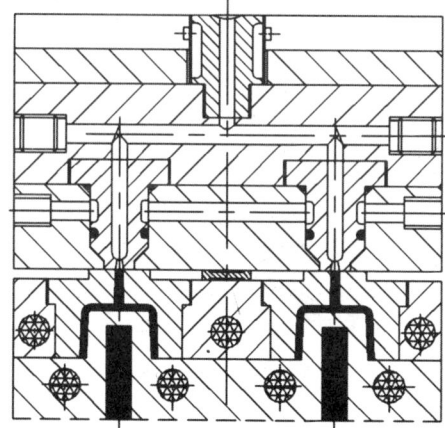

Fig. 14.7

thermosets are also available in various sizes, with open or shut off nozzles.

Figure 14.8 shows one such system from a German manufacturer of hot runner devices. It can have open nozzles as shown on the left or the valve gated ones as shown on the right side. The system is water cooled. The needle valves are actuated pneumatically.

The Gate

For single cavity moulds, a sprue gate is quite suitable if the gate mark is not found disturbing. It causes very little pressure loss and fills the cavity very fast with less injection pressure. The sprue gate is, however, not suitable for aminoplasts as it may give rise to internal stresses around the gate area.

The best cross section for a gate is a circle followed by a square. All types of conventional gates, including the submarine gate, can be used for thermosets. Most common are the film gate and the edge gate. The film gate is thin, is easy to remove and leaves a very small mark on the moulding. It causes very little orientation and does not damage the fibres. The edge gate is easy to make and modify but requires some post operational work for neat removal.

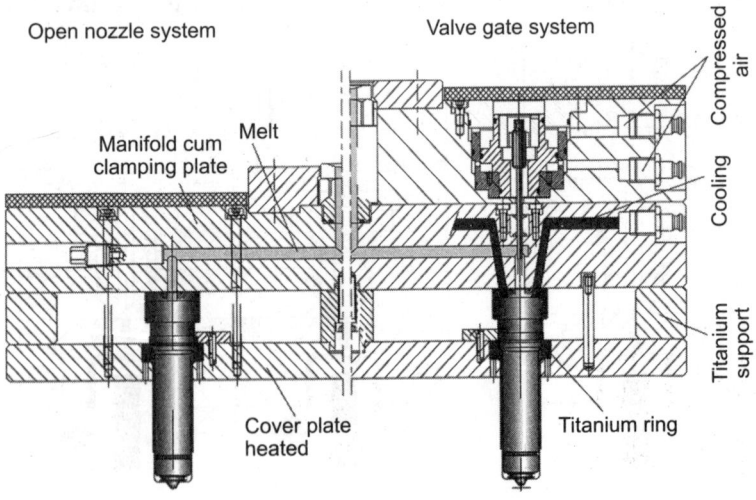

Fig. 14.8

A modified version of the film gate (Fig. 14.9), which consists of a number of edge gates from the main runner in place of a continuous film, provides the possibility of straightening the flow front by selectively altering the size of individual gates. It proves particularly advantageous with relatively flat rectangular articles as the melt can be made to reach the other end simultaneously at all points. In order to keep the flow path as short as possible, the film gate should be situated on the longer side of the moulding.

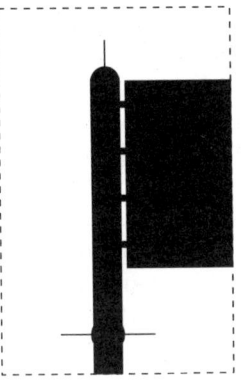

Fig. 14.9

The submarine gate obviates post-operations but calls for meticulous adherence to rules concerning its shape, size, inclination and the placement of ejectors for safe demoulding. These factors are governed by the nature and composition of the compound. It can, however, damage the fibres if the gate is too small.

The side gating of a cylindrical moulding may bring forth a poor weld line as the flow fronts may be in advanced state of cure as they meet. A diaphragm gate may be more suitable here.

The gate for materials with fine fillers should have a depth of 0.5 mm. or more and that for coarse ones between 0.75 and 1 mm to start with. Too small a gate can also result in abnormal increase in temperature, bringing the outer skin to an advanced stage of cure. It would not bond with the following melt.

The filled thermosets, passing through narrow gates at very high speeds, cause rapid wear in the gate area. The gate sections should, therefore, be designed as replaceable inserts made of hard metals such as carbides.

The gate location should take into account the usual problems of weld lines, product appearance, venting, cost of gate removal etc.

The ideal location for the gate is the thickest section of the moulding. The free flow of the melt in case of thermosets does not cause problems like those encountered with thermoplastics.

The arrangement of cavities in a multi-cavity mould should ensure equal flow paths for the material from the central source if it does not result in extraordinary increase in length of the runner. Experience has, however, shown that the balanced runner system is not always necessary to produce good mouldings if the product is relatively small. Here, the runner system, being small, fills first and the built up pressure causes the melt to enter all cavities simultaneously. However, in case of large articles, the cavities closer to the main sprue will start filling earlier than the remote ones. A balance can be achieved by varying the gate size.

The Ejection System

In principle, the ejection devices in injection moulds for thermosets are no different than those for thermoplastics. However, the resin extruding down the ejector pin holes poses additional problems in moulds for the former. It leads to seizure of the ejection assembly. The situation gets worse with progressive wear. To counter the hazard, the ejector pins over 5 mm diameter are provided with flats (Fig. 14.10). A triangular cross-section after a smooth length equalling one diameter in the front offers maximum of alignment with minimum of friction and wear. The flats should extend well enough along the length of the pin so that the escaped resin and other foreign matter may fall away freely. The flats must terminate at the moulding end of the pin with a face at right angles to the surface of the flat. This helps to break up and fragment any resin that extrudes down the pin, to fall away instead of forming a tube around the pin. The flats should leave 1,5 mm broad guiding surface along the shank. For pins smaller than 5 mm in diameter, the breadth of flats may be less but their number should be at least three. It is, however, advisable not to use ejectors less than 6 mm in diameter.

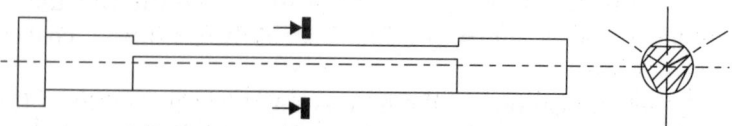

Fig. 14.10

Almost all devices of clearing undercuts, employed in injection moulds, are applicable here too. Internal undercuts, however, cannot be circumvented by forced ejection. The air jets should ensure that no chips or material fragment remain sticking in the guides for slides.

The Mould Venting

The moulds being filled in closed condition need venting on the parting line and in deep crevices to allow the hot air or the volatiles, released by the compound while entering the mould, to escape.

The usual technique for venting is to grind "flats", about 0.07 mm deep and 6–10 mm broad on the face or land of the cavity in locations where the melt would reach last. It is, however, not always possible to predict the flow pattern. In such cases, the vents are ground in the mould after the first trial. Lack of proper venting may lead to surface porosity, incomplete filling, burn marks and staining of the mould in areas where the volatiles have not been able to escape. The venting slots should be highly polished to reduce the risk of flash sticking onto the mould.

The air trapped in deep sections such as bosses, ribs, etc cannot be vented by the slots on the parting line. Ejectors and inserts placed at the strategic points can be a good solution in such cases.

The technical components moulded out of thermosets may not require very high surface finish but a rough mould surface acts against smooth ejection. The vertical walls of the mould cavity should, therefore, be well polished in direction of ejection.

The Mould Material

As the injection moulds for thermosets operate at elevated temperatures, it is advisable to fabricate mould plates and core and cavity inserts out of hot die steels like DIN 1.2343 which retain their strength at those temperatures. Standard mould components like ejector pins are available in heat resistant steels. The steels should be polishable.

The gate insert is subjected to considerable friction and wear. It must be made of a through hardening steel and for filled materials out of carbides.

The Mould Shrinkage

Thermosets also undergo shrinkage. The shrinkage is measured 24–168 days after moulding and compared with the tool dimensions at 20 ± 2°C. The shrinkage is usually specified by the raw material manufacturer and it is added to the product dimensions proportionately while determining the mould dimensions. The factors contributing to shrinkage are:

- Volume contraction due to curing
- Escape of moisture from the resin
- Escape of volatile components
- Difference of thermal expansion between the mould and the moulding compound.
- Elastic relaxation of the material after release of pressure.
- The shrinkage increases with increasing mould temperature. It may be attributed to the differential thermal expansion between the tool and the moulding material. Secondly, the material loses more volatile matter at higher temperatures.
- The shrinkage decreases with increasing injection and holding pressures. An increase in holding time results in significant decrease in shrinkage. The effect is reverse in case of the injection speed.
- Shrinkage of materials impregnated with fibrous matter is not uniform in all directions. It is greater in the radial direction than in the tangential one, due to orientation of fibres. The ratio may be as high as 2:1. Pure resins exhibit negligible difference in shrinkage in different directions.

The Post-shrinkage

The moulded articles undergo further contraction upon conditioning after cooling. It is carried out for 168 hours in an oven at 110°C. The cause may be reorientation of molecules as well as further loss of volatile constituents. It follows the pattern of the main shrinkage, viz. more in the radial direction than in the tangential one.

THE PRODUCT DESIGN

The rules for design of articles produced by casting of molten plastics under pressure are almost universal. Guidelines valid for the design of moulded thermoplastics products also apply to those out of thermosets. There are, however, variations in some cases due to peculiarities of materials and their processes. The following points merit special attention:

Undercuts: External undercuts usually require devices like slides for clearance. Moving mould components function well if their guides are kept free from the dirt, loose material chips, powder and flash particles. The moving devices are costly in any case. If possible, the undercuts should be avoided or formed in a way that permits ejection without slides.

Parts with internal undercuts like continuous grooves and beads, etc., unless very shallow and well rounded, cannot be ejected by "jumping over" the obstruction after injection moulding because of the rigidity of the material. It is also advisable to avoid all internal undercuts requiring parting of the core as far as possible.

Although it is possible to incorporate slides and other undercut clearing devices in case of injection moulding, it must be borne in mind that the fillers will part around the obstruction and form a weld behind it, which may prove to be a weak point.

Wall thickness: Regardless of the compound employed, the wall thickness should be kept uniform and to a minimum, consistent with the functional requirement. It determines the curing time, which in turn controls the overall cycle time, productivity and the cost. Unequal wall thickness results in variations in curing time of different sections and leads to non-uniform density. Very thin sections may hinder the flow of fillers and the filler-less compound may cure there much faster. All these factors effect the ultimate mechanical, electrical and chemical properties of the end product. Table 14.1 gives minimum and maximum recommended wall thickness for various materials.

Ribs: Large flat surfaces tend to distort. Ribs lend stiffness, help in saving material and shortening curing time and also work against warpage. Their wall thickness should be

Table 14.1: Recommended wall thickness		
Material	*Wall Thickness mm*	
	Minimum	*Maximum*
Phenolics, General purpose	1.25	25
Phenolics, flock filled	1.25	25
Phenolics, glass filled	0.8	19
Phenolics, fabric filled	1.6	9.5
Phenolics, mineral filled	6.5	25
Urea, cellulose filled	0.9	4.5
Melamine, cellulose filled	0.9	4.5
Epoxy	0.8	25
Polyester	1.0	25

preferably less than that of the article; in no case should it exceed the later. A taper of at least 1° on each side, or more if feasible, helps in safe ejection. Their height too should not be more than three times their thickness. Generous fillet radii ensure strength, easy filling and ejection. Flow lines may appear on the back of ribs in injection moulding. These can be eliminated or camouflaged by introducing shallow flutes or decorative features (Fig. 14.11).

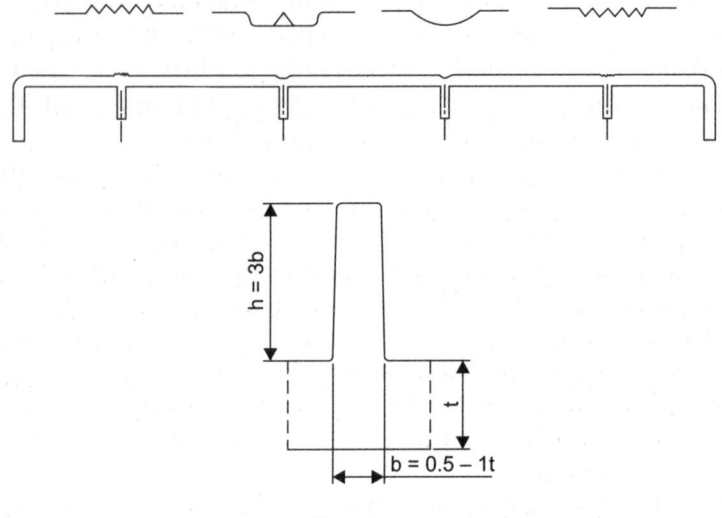

Fig. 14.11

Bosses: The wall thickness of a boss should not exceed that of the wall on which it stands. Its joint with the adjacent side walls should not result in material accumulation. It may be joined to the side wall by a rib (Fig. 14.12).

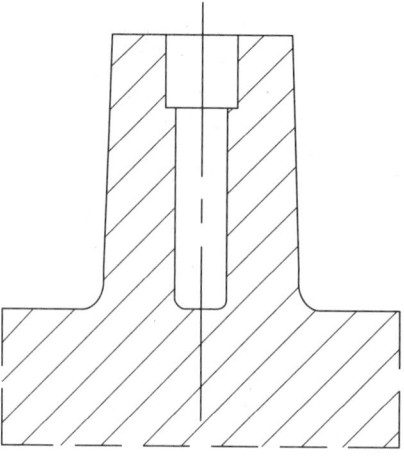

Fig. 14.12

Corner Radii: Sharp inner corners retard the flow of material. During curing and shrinking, the inner corners may generate stresses. Sharp corner may give rise to cracks.

As a rule, an inner radius of 0.75 times the wall thickness proves adequate. It has been found that the larger radii do not influence the stress situation significantly. They, however, ease the flow.

Draft: The injection moulded parts are generally demoulded with the help of ejector pins, which are placed close to, but not directly in line with, the forces of retention. Unless the sidewalls, on which the article shrinks, are foreseen with adequate taper, the moulding may get damaged or distorted during ejection. A minimum taper of 0.5° is a must, though a larger angle would be safer.

Threads: It is possible to form male or female threads in injection moulded parts, usually in a semi-automatic mode by placing threaded cores in the mould, ejecting them along with the moulding and unscrewing them outside. Threads with v-profile are most suitable. The pitch, however, should not be less than 0.8 mm, especially for compounds with coarse fillers. Both the start and the end of the threads should be about 1 mm below the surface as shown in Fig. 14.13. Threads can also be cut after moulding. The moulded core hole should be conical to avoid excessive friction while tapping. If drilled, the core hole should be bigger by 0.5° mm than that for metals.

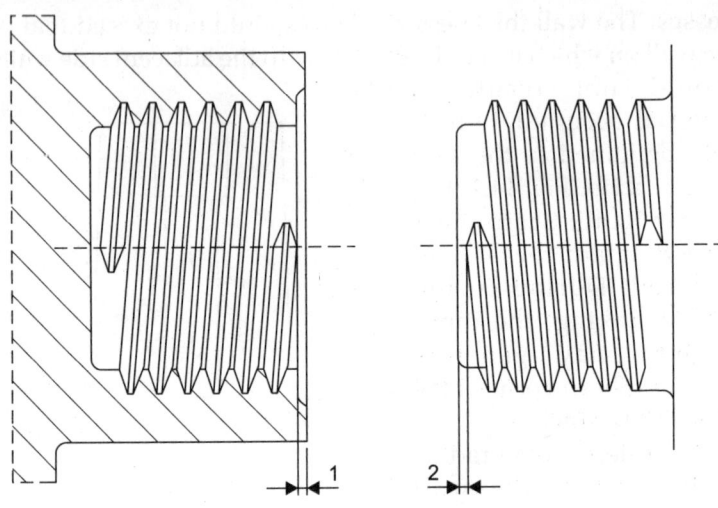

Fig. 14.13

Screws: Parts moulded in thermosets can be fastened together by means of screw fasteners. A cylindrical head screw is preferable to one with a countersunk head as wedge action of the later may lead to cracks and fracture (Fig. 14.14).

Inserts: Metal inserts, placed in the mould before moulding, can save certain post-operations like drilling and tapping. The inserts are foreseen with anchoring aids like diamond knurling on the outer surface. They should not have sharp edges, which could cause cracks. Also electrical contacts and assembly lugs etc. can be moulded as inserts. Inserts with special configurations can also be pressed in right after ejection when the moulding is still hot.

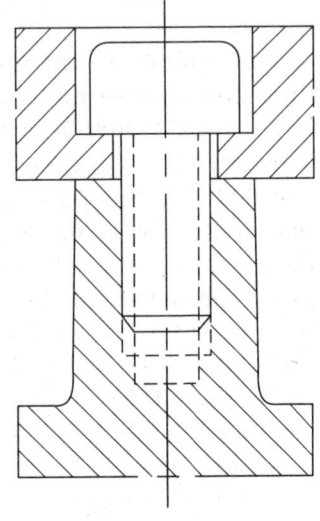

Fig. 14.14

TROUBLE SHOOTING
(Moulding Defects, their Causes and Remedies)

A number of factors influence the quality of a moulding. A flaw or incongruence there results in a defective product. Recognition of the cause is the first step towards finding the remedy. The influencing factors are:

1. The properties of the material
2. The product design
3. The screw geometry
4. The screw speed
5. The mould temperature
6. The injection pressure
7. The holding pressure
8. The curing time

The Flash

A thin material film at the parting line, joined with the moulding.

Causes

- The machine locking force insufficient.
- Too high an injection pressure.
- Too late switchover from injection to holding pressure
- Mould temperature too high; the curing is too fast
- Mould parting not parallel or damaged

Remedies

- Shift to a bigger machine
- Reduce injection pressure/speed
- Switch over to holding pressure earlier
- Reduce mould temperature
- Grind/repair the parting surface,

Demoulding Problems

The part sticks in the mould and distorts during ejection.

Causes

- Draft too small
- Mould surface too rough.

- Ejectors insufficient, too small in area, unbalanced positioning.
- Holding pressure too high and/or too long.
- Mould temperature too high
- Part not fully cured.

Remedies

- Increase taper if possible
- Check and improve mould surface
- Replace ejectors, increase size, add more ejectors
- Reduce pressure/ pressure duration
- Reduce mould temperature
- Increase pre-heating time

Dull Areas

Dull zones on the surface.

Causes

- Sediments on mould walls, release agent.
- Local hot spots in the mould causing colour variation.
- Porosity, caused by volatiles, incomplete filling, air entrapment.
- Impurities in the moulding compound.
- Low moulding pressure.

Remedies

- Clean the mould.
- Equalise mould heating.
- Preheat the compound.
- Improve venting
- Increase holding pressure/time.

Blistering

Hollow high spots on the surface of the moulding.

Causes

- Gases, trapped under the hardened skin, lifting up the layer.
- Mould temperature not uniform.
- Inefficient venting.
- Product not fully cured.

Remedies

- Increase preheating to drive away the volatiles.
- Equalise mould temperature.
- Improve venting.
- Introduce breathing.
- Increase curing time.

Orange Peel

Pit marks, elevations and depressions on the surface.

Causes

- Coarse/ hard grain compound. Skin hardens but coarse grain below cures later and unequally.
- Flow of material too slow, grains not melted.
- Mould temperature too low.

Remedies

- Change material, use refined/ softer grain.
- Decrease preheating time.
- Increase mould temperature.
- Increase pressure.
- Increase curing time.

Sink Marks

Shallow depressions on the surface, visible through gloss difference.

Causes

- Material accumulation.
- Unfavourable rib-to-wall thickness ratio.
- Fillet radii too big.
- Vent channels too deep.
- Curing time too short.
- Moulding pressure too low.

Remedies

- Change product design, reduce material accumulation.
- Correct venting channels.

- Increase the material dosage.
- Increase pressure.
- Increase curing time.

White Areas

Some areas on the surface appear light-coloured (applicable to UF, MF).

Causes

- Thermal degradation of pigments.
- Barrel temperature and/ dwell time too high.
- Mould temperature too high at some spots.
- Material not flowing easily.

Remedies

- Decrease barrel temperature.
- Increase injection speed.
- Decrease overall cycle time.

Silver Streaks

Elongated bubbles, silvery in appearance, on the surface.

Causes

- Moisture in the compound forming steam.
- Venting ineffective/ clogged.
- Mould closing too tight.

Remedies

- Change the compound.
- Facilitate escape of moisture by injecting slowly.
- Clean/ improve the vents.

Warpage

The part deforming after ejection.

Causes

- Incomplete curing due to low mould temperature.
- Insufficient curing time.

Remedies

- Increase mould temperature.
- Increase curing time.

Diesel Effect

Burn marks on the product.

Causes

- Too much lubricant in the compound.
- Entrapped gases, poor venting.
- Choked vent holes
- Mould temperature too high

Remedies

- Change compound.
- Reduce preheating
- Reduce mould temperature.
- Improve venting.
- Change gate position.
- Clean vent slots, ejectors.

Cracks

Through-going crevices especially near sharp corners, cut-outs, change of section, inserts.

Causes

- Internal stresses caused by obstructed shrinkage near sharp corners.
- Plastics layer around inserts too thin.
- External stresses during ejection.

Remedies

- Round off sharp corners.
- Preheat metal inserts.
- Redesign product with more material around inserts.
- Decrease mould temperature.
- Increase curing time.
- Increase demoulding draft.
- Place ejectors at crucial points

ADVANTAGES AND DISADVANTAGES
OF INJECTION MOULDING

Advantages

- Homogeneity due to mixing action of the screw.
- Greater accuracy
- More freedom in product design as wall thickness may vary.
- Less post-operations
- More uniform temperature of the melt in the mould, consequently more uniform curing and more uniform shrinkage. Closer tolerances.
- Shorter cycles
- Automation

Disadvantages

- Higher investment
- Greater orientation of fibres
- Special material formulations

15

Injection Moulding of Liquid Silicone Rubber

- The material
- The moulding equipment
- Two component moulding
- Applications
- The moulding process
- The mould
- Trouble shooting

THE MATERIAL

Liquid silicone rubbers are high purity platinum cured silicones. They belong to the family of thermosetting elastomers that have a backbone of alternating silicone and oxygen atoms and methyl or vinyl groups and vulcanise by the process of polyaddition. The reaction starts when two components of the material are mixed together. The speed of the vulcanisation is proportional to the ambient temperature. The reaction is very slow at room temperature but proceeds very fast at temperatures above 150°C. This characteristics of the liquid silicone rubbers forms the basis of their processing by injection moulding, which may be considered similar to that for the thermosetting polymers (see Chapter 14). In both cases, the material is injected in fluid, uncured state into the mould, where the temperature initiates chemical reaction resulting in hardening of the melt.

Liquid silicone rubbers possess following characteristics:
- High transparency
- Low compression set
- Flexibility between –50°C and +200°C
- Light resistance (including UV and X-rays)
- Resistance to steam
- Very good ageing resistance
- Very good long term mechanical endurance
- Constant mechanical properties over a wide temperature range

- Excellent electrical insulation
- Excellent environmental compatibility
- No toxic products on burning
- Adhesion to metals and plastics possible with additives
- Repulsion to water

THE MOULDING PROCESS (FIG. 15.1)

The liquid silicone rubber for injection moulding is supplied, ready for use, as two separate fluids, usually referred to as component A and component B. One of the component contains the platinum-based catalyst. These are stored separately and pumped under pressure in equal proportion to a static mixer. It is also possible to include a pigment conveyor in parallel connection to the mixer. These devices are maintained at a controlled temperature of 18–20°C. The cold mixture is conveyed at a pressure of 30–70 bar to the injection unit of the injection moulding machine, which is similar to that of an injection moulding machine for thermosets. The barrel differs from its counterpart in conventional moulding in two aspects; it has no hopper and no heaters. The moulding material in the form of a liquid mixture comes to its feeding section through a flexible pipe from the static mixer and it is provided with a cooling jacket for circulation of a cooling fluid, usually water. The screw in the barrel has two functions; it acts as a conveyor of the fluid mixture from the feeding point to the machine nozzle at the end of the barrel and as a piston to inject the fluid into the mould. The material transport takes place under a back pressure of 5–30 bar. The screw does not have to compress

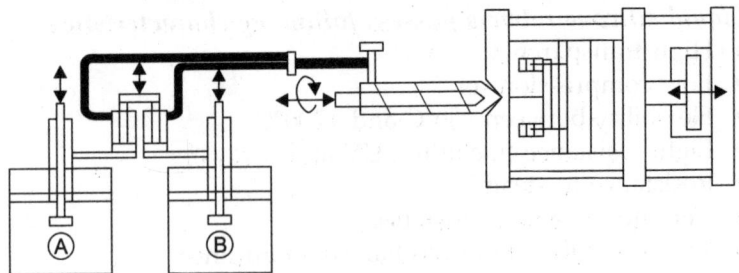

Fig. 15.1

and plasticise the mixture which is already in a fluid state. It is also not necessary to make the fluid travel a certain distance from feeding point to the nozzle for plasticising. Therefore, the feeding zone in converted moulding machines is usually advanced in position. The cooling arrangement of the injection unit is tailored to maintain a uniform temperature between 18–20°C.

Melt Injection

The operation of injection is similar to that for thermosets. The melt is forced into the mould, which is maintained at an elevated temperature, by the screw acting as a piston moving linearly. The uncured liquid silicone rubbers are very thin flowing and hence do not need high injection pressures. Most varieties can be moulded with an injection pressure close to 100 bar. The material cushion is kept to a bare minimum, almost zero, in order to avoid flashing. The silicone rubbers expand in volume on curing, thereby increasing the cavity pressure manifold, in some cases upto 300 bar. To take care of the expansion, the cavity is filled to about 95–98 % by injection and gets filled completely through increase in volume of the crosslinking silicone rubber.

Testing of a new mould invariably starts with intentional short shots. The optimum filling limit is explored through a number of progressive shots to avoid overloading and flashing. The speed and feed is chosen to fill the mould as fast as possible to forestall curing. The injection speed is throttled towards the end of injection stroke to enable effective venting. The vulcanising or the curing time, which takes the place of the "cooling time", depends upon the wall thickness of the moulding and ranges between 3–5 seconds per mm. Vulcanisation may be considered to be over when the article can be ejected without damage or permanent distortion.

Unlike thermosetting polymers, silicone rubber does not undergo any contraction during curing in the mould; it expands on the other hand and tends to ooze out of the gate after termination of injection. Holding pressure, unlike with thermoplastics moulding, has the sole function of preventing the reverse flow till the gate section has cured and closed. It may take about 1–4 seconds, depending upon the size and location of the gate. The holding time can be considerably shorter with shut-off nozzles. The holding pressure may range between 50–100 bar.

A measure specific to the moulding of LSR is the so-called "bumping", which is carried out to dislodge entrapped air bubbles. It is sudden release of moulding pressure followed by its build-up to full extent. It is repeated several times in rapid succession during injection before the curing process sets in.

EJECTION

Although the moulding machine is equipped with the usual ejection device and the mould also has ejectors and strippers, the flexibility of the cured moulding makes the use of additional external aids unavoidable. These may be rollers, pickers, robots, etc as discussed later.

Interrupting or Terminating the Moulding Process

The polymer starts crosslinking right from the moment of mixing of the two components. Its speed of reaction, however, is very low at 18–20°C at which the mixer, the flow lines and the injection unit of the moulding machine are maintained. Short breaks, lasting a few hours, do not cause much damage but an interruption involving more than three days calls for special measures. It becomes necessary to either purge the injection unit with the component A or store it at –20°C. The same treatment must be meted out to the cold runner assembly of the mould.

If the production is required to be stopped for a longer period, all machine and mould parts containing the mixture must be flushed out with the component A of the polymer. Storage of the above parts at –20°C obviates purging.

In locations with high or fluctuating temperatures, air conditioning is necessary for trouble free production.

Post-operations

LSR mouldings are generally ready for use after cooling but post curing is carried out in special cases like those where foodstuff or medicines are involved. It is also resorted to for obtaining a low compression set. The process involves baking the mouldings in a forced circulation oven at an elevated temperature (~200°C) for a certain period (4 hours for 2 mm. thick plane parallel components) to drive out volatile constituents, which are primarily combustible siloxanes. Clean fresh air is passed through constantly to avoid deflagration. Good ventilation of the site is imperative.

THE MOULDING MACHINE

The injection moulding machine for processing liquid silicone rubbers differs from its counterpart for thermosets only in a few design features of the plasticising unit. In the first instance, it cannot be truly designated as a plasticising unit as it receives the raw material in a fluid state which is not altered in the barrel in any way till the point of injection. The unit does have a barrel, a screw, a non-return valve and a nozzle. The screw does not have to compress and plasticise the material. Its function is limited to that of a conveyor and a piston. The conventional design of the non-return valve is inadequate for the extremely thin flowing liquid silicone rubber. Special valve with spring support has been devised for this polymer. The machine nozzle too must be of the shut-off type, preferably a hydraulically operated one. The barrel does not have a hopper and heaters. It is provided with a cooling jacket to keep its temperature within a narrow range of 18–20°C. The machine nozzle is also water cooled. As the material is not required to traverse a certain distance for its plastication, the feeding zone is shifted further towards the exit end.

The mould closing and ejection units do not differ from those of the conventional injection moulding machines in any way. It is, therefore, possible to employ a conventional machine for processing of LSR by changing only the injection unit. The very low viscosity of the mixed fluid, however, makes very fine control of the injection stroke and the injection speed imperative. It is, in the end, economical to employ a closed loop machines with precise controls for pressures and speeds.

A prime requirement of the closing unit is the parallelism of the highest order between the mould platen to prevent flashing of the thin fluid. The injection pressure needed for processing of LSR is much lower than that for thermoplastics. It enables moulding of articles with large surface areas on machines with lower locking force.

Ejection of flexible LSR mouldings only with conventional devices like ejector pins, sleeves and strippers is not always successful. Additional supportive arrangements like roller sweeps, brushes and pickers add a measure of certainty to the operation of demoulding. Robots are a useful accessory to a moulding unit for LSR.

Controls for the mould heating may be separate in case of converted machines but machines devoted specifically to LSR moulding have them integrated in their systems. Same holds good for the vacuum devices for mould venting. In order to facilitate evacuation of air from the mould, these machine have the arrangement to close the mould with a low pressure, inject the material and then apply the final locking force. Another arrangement to enable creation of vacuum in moulds with a seal on their parting face is to stop the moving platen 0.5–2.0 mm. before the mould halves touch each other, suck out the air and then lock the mould before injecting the mixture.

A moulding machine for a particular product is selected on the basis of the locking force needed to counterbalance the cavity pressure generated by the expanding polymer injected in the mould. To be on the safer side, a cavity pressure of 400 bar per square centimeter is assumed. The locking force of the machine should be slightly higher than the total cavity pressure. Another feature, no less important, is the shot weight. The screw must make a stroke equalling 1–5 times its diameter to inject the shot otherwise its revolutions for dosing the next charge would be insufficient for an adequate mixing. The injection unit should not be oversized.

THE MOULD

The Mould Design and Construction

In principle, an injection mould for LSR is almost identical to its counterpart for injection moulding of thermoplastics. It is required to perform the same basic functions and has the same or similar design features, viz. a feeding path for the "melt" from the entry point to the cavity, the cooling (here the heating) system for solidification of the injected fluid and an arrangement to eject the mouldings without damage and distortion.

The extraordinary high fluidity of LSR puts extreme demands on the leakproofness of the mould. Any deviation in parallelism of plates may prove disastrous. Smallest distortion and gaps to the tune of 0.01–0.02 mm. result in flash. The best way to eliminate distortion caused by machining stresses is to anneal the mould plates after every machining operation.

The feeding system can be similar to that of thermoplastics but the cross-section of the runners, which should preferably be circular, may be much smaller. The runner system must be a balanced one to fill all cavities evenly.

Conventional gates such as edge gate, film gate, pin point and submarine gate are also applicable for moulding of silicone rubbers though their size can be much smaller. For instance, a submarine gate with a diameter of 0.2–0.5 mm. proves adequate in most cases.

Conventional feeding systems with primary and secondary runners result in considerable wastage as the runners cannot be recycled. To circumvent this drawback, a cold runner system, similar to the one for moulds for thermosets (chapter 14), has also been developed for moulding of LSR. The aim is to bring the polymer mixture from the machine nozzle up to the mould cavity in uncured state, i.e. without any appreciable rise in temperature. The cold runner system is housed in a separate block, integrated in the mould but insulated from it. An efficient cooling network is essential to prevent a rise in its temperature through conduction, convection and radiation from the hot mould surrounding it. The nozzles of the cold runner system, bringing the polymer into the cavity, are equipped with shut-off needles, operated hydraulically or pneumatically, to prevent drooling of the mixture when the mould opens.

Multi-point gating is, in the rule, not necessary even for large articles due to excellent flowability of the "melt". More than one gate may introduce another problem too. Even minute differences in the gate size may cause unbalanced filling.

The necessity for rapid injection to forestall premature curing before complete filling is accompanied by difficulties in efficient expulsion of air out of the closed mould. Venting is usually provided on the mould parting line in the form of shallow slots. Air vents deeper than 0.004–0.005 mm. will give rise to undesired flash. The slots are usually 1–3 mm. wide and are located at the point of last filling. When compressed highly, the trapped air gets overheated and causes whitening of edges due to degradation of the compound.

To evacuate enclosed air from a critical place, the parting line may be made to pass through it.

In critical cases, when conventional venting proves inadequate, it becomes necessary to evacuate air from the mould before injection of material. To this end, the mould is provided with a flexible, heat resistant seal around the cavity on the parting face. The mould is closed incompletely, leaving a gap of 0.5–2 mm. between the mould halves. The enclosed air is sucked out with the help of a vacuum pump and the mould is closed fully before the injection takes place.

Moulds for LSR, particularly the plates housing the cavities and cores, must have a uniform temperature. Depending upon the polymer, these mould components are maintained at a temperature between 170–220°C, usually through incorporation of cartridge heaters. A thumb rule for determining the heating capacity is allocating 500 watts for every kilogramme of steel to be heated. Oil heating is preferred for large moulds. In order to preclude any distortion, the holes for cartridges or for oil circulation should not be too close to the parting line of the mould. If only core and cavity plates are being heated, these must be insulated from the rest of the mould bolster through insulating plates.

The Ejection System

Liquid solid rubbers do not shrink in the mould, they expand upon crosslinking. The moulding may not remain on the core side when the mould opens. The sticking tendency of the LSR moulding may make it adhere to the mould side with greater surface area. In ambiguous cases, it is more practical to determine this in a trial mould before going in for a production tool. One way of keeping the mouldings on the desired side is to choose parting line forming an undercut. The flexibility of the material enables demoulding without special arrangements of clearing undercuts.

Among the ejection devices, the cylindrical ejector pins are the least effective. They tend to pierce the flexible moulding sticking to the mould surface. They may be employed selectively in conjunction with other internal and external ejection devices such as :

- Strippers
- Sleeve ejectors
- Mushroom ejectors
- Air ejectors
- Roller sweep
- External brushes and pickers
- Robots

The Mould Shrinkage

As mentioned before, liquid silicone rubbers expand during curing instead of shrinking. Shrinkage takes place while cooling after ejection and varies between 2.5–3%. The extent of shrinkage depends upon:

- Mould temperature and demoulding temperature
- Pressure in the cavity
- Gate location; shrinkage is more along the flow and less across it.
- Wall thickness; thicker sections shrink less
- Post curing; additional shrinkage of 0.5–0.7%.

Mould Materials

Steels used for the mould components must retain their hardness, strength, wear resistance, parallelism and polish at temperatures above 200°C. The mould bolster may be fabricated out of an unalloyed steel like EN8 but the mould plates must be made of a tougher material such as P20. The heated plates and mould inserts must be formed out of hot die steels. AISI H11 is highly suited for this application. The filled varieties of LSR cause a considerable wear in the mould. Ti-Ni treatment of the mould surface prolongs the mould life considerably.

The mould surface should not be polished too much as SLR tends to stick on shining surfaces. If permissible, the surface should be roughened either through spark erosion, etching or by wet sandblasting. The recommended surface roughness is 0.5–3.2 μm. Only in case of transparent mouldings, there is no alternative to polishing the mould. The sticking tendency of the material can be countered by a PTFE-Ni treatment of the mould surface.

The mould must be provided with insulating plates on both sides to prevent loss of heat to the machine platen.

TWO COMPONENT INJECTION MOULDING WITH LSR

It is possible to mould liquid silicone rubbers over one another as well as over thermoplastics and metals. As a rule, these rubbers do not bond with thermoplastics and metals. The bond has to be achieved through mechanical means like undercuts or by application of primers over thermoplastics and metal substrates. However, it is also possible to compound the bonding agent with the LSR so that the process of overmoulding can be carried out without interruption. The adhesion improves with ageing and reaches the optimal level in about two weeks after moulding. The process can be accelerated by heat treatment to 100°C for 1 hour.

There are two ways to perform the two component moulding:

Tandem Moulding

The method employs two injection moulding machines, running side by side. One of the machines produces thermoplastics substrates which are transferred to the second machine devoted to the moulding of LSR. A robot positions the mouldings from the first machine in the mould on the second one. After overmoulding, a composite article is ejected out.

The process needs two standard injection moulding units as well as a handling device. Investment in capital and space is, of course, high. The cycles, too, have to be synchronised.

The advantage lies in the simplicity of moulds and independence of the choice of working parameters from each other, which can be optimised according to the need of the process and the material in each case.

Composite Moulding

The process requires a special injection moulding machine with two injection units. The unit for the thermoplastics moulding is heated whereas the one for LSR is maintained at temperatures around 20°C. The mould houses equal number of cavities for the thermoplastics substrate and for the final moulding. As the mould opens after moulding of the thermoplastics substrates, the core plate of the mould is rotated by 180 degrees to position the substrate mouldings opposite the overmoulding cavities. Simultaneously, the other set of empty cores gets positioned opposite the corresponding cavities for injection with thermoplastics. The mechanism for rotation of the core plate is housed either in the mould itself or the machine has a built-in device to rotate the moving mould half after each shot. The mould may also contain a hot runner system for thermoplastics and a cold runner system for the liquid silicone rubber. A compromise has to be made about the mould temperature, which as a rule, should be much higher for the LSR than for the thermoplastics. As a way out, usually those thermoplastics are used for substrate which permit higher demoulding temperatures. These are: polyamides, polyacetals, polycarbonate, polyester, polyphenylene sulphide, etc. The liquid silicone rubbers too are of the type which vulcanise at lower temperatures. The mould is heated to about 110–130°C.

As obvious, the single machine alternative is economical in capital cost and space requirement. The compromise between conflicting processing requirements of the two different materials usually results in slower production. The mould, too, becomes highly complicated.

Two component mouldings may also be made out of two different types of LSR for having two different colours or for making one side or one part of the product electrically conductive.

TROUBLE SHOOTING

Moulding defects, their causes and remedies.

Short Moulding

The moulding is not filled completely/has less weight/has wavy surface.

Causes

- Injection speed/pressure too low
- Material flowing back
- Dosage insufficient
- Inadequate venting
- Premature curing
- Back leakage in the barrel
- Runners/gate too small
- Runner system unbalanced

Remedies

- Increase injection speed/pressure
- Delay switch-over to holding pressure
- Increase shot volume
- Increase holding time
- Improve venting
- Reduce mould temperature/barrel temperature
- Check non-return valve
- Increase runner/gate size
- Balance runner system

Burn Marks

White edges, streaks, blisters

Causes

- Overheating of the mixture
- Inadequate venting
- Very high injection speed
- Very high mould temperature

Remedies

- Decrease mould temperature
- Decrease injection speed
- Improve mould venting

Orange Peel
Pit marks, depressions and raised spots on surface.

Causes
- Injection speed too low
- Mould temperature too high

Remedies
- Raise injection speed/pressure
- Decrease mould temperature and cold runner temperature

Included Air
Bubbles in the moulding.

Causes
- Mould temperature too high
- Cold runner temperature too high
- Injection speed too low
- Inadequate venting (if bubbles are away from the gate)

Remedies
- Reduce mould temperature
- Reduce cold runner temperature
- Increase injection speed (gradually)
- Improve venting (if bubbles are away from the gate)

Demoulding Problems
Article sticks in the cavity, damage during ejection

Causes
- Mould too hot
- Holding pressure too high/too long
- Curing time too long
- Mould surface too glossy

Remedies
- Decrease mould temperature
- Reduce curing time
- Roughen up mould surface/coat with PTFE

Flash
Silicone film at parting line.

Causes

- Shot too large
- Injection speed too high
- Switch-over to holding phase too late
- Clamping force too low
- Mould plates not plane/damaged
- Venting slots too deep

Remedies

- Reduce shot volume
- Reduce injection speed
- Set switch-over point earlier
- Increase clamping force
- Change over to bigger machine
- Check mould plates and rework
- Reduce depth of vents.

Article Size Incorrect

Dimensions are out of range.

CAUSES

- Insufficient curing.
- Different shrinkage
- Deformation on ejection

Remedies

- Increase curing time
- Check mould temperature and increase if necessary
- Check for uniform mould temperature
- Improve ejection, roughen cavity surface

Tacky Surface

Greasy areas on surface.

Causes

- Incomplete curing
- Mould temperature too low
- Components' proportion unequal
- Inhibitors in vicinity

Remedies

- Increase curing period
- Increase mould temperature
- Check heaters and thermocouples
- Check mixer for pressure fluctuations
- Clean mixer and supply lines
- Check for sulphur compounds, rubbers, urethane and epoxy resins, etc. in neighbourhood and remove

FIELDS OF APPLICATION

- High temperature gaskets and seals for automobiles, household gadgets, food processing machinery, electrical and electronic equipment, connectors
- Pump membranes and components, bellows
- Medical appliances, inhalation masks
- Kitchen goods, baking pans, spatulas
- Diving masks
- Nipples for baby feeding bottles

Reaction Injection Mouldings

- The process
- The equipment
- Applications
- The materials
- The mould
- Advantages

THE PROCESS

RIM is a process for rapid production of plastics articles directly from low viscosity monomers. Two or more liquids are combined by impingement mixing at high pressure and then fed into a mould at very low pressure. Solid polymers are formed by cross-linking of the monomers in the mould. RIM is, therefore, a chemical as well as a moulding process and quite different from conventional thermoplastics injection moulding because it uses polymerisation, rather than cooling, to form a solid polymer which has properties different from those of the materials being fed into the mould. Although moulding processes like monomer casting or thermoset injection moulding also use polymerisation to set the part shape, they employ hot mould walls to activate the reaction. In RIM, monomer and mould temperatures are not so different and the reaction is initiated by the impingement mixing.

Reaction injection moulding process differs from the classical injection moulding in many of its features. The plasticising does not take place in a barrel through heat and mixing and injection is not achieved by means of a screw. The "melt" does not need high pressure for injection into the mould. The filling into the cavity is more like casting than injection. Again, the solidification of the "melt" into the moulding takes place neither by extraction of heat, viz. cooling, nor by addition of heat as with the moulding of thermosets. The polymerisation of the injected mixture of monomers leads to solidification. The mould plays the role of a container for polymerisation.

In case of polyurethane, two or more low viscosity (0.1–1.0 Pa.s) liquid reactants (di-or trialcohols and di- or tri-isocynate), which are held in separate temperature controlled tanks equipped with agitators, are fed through supply lines to metering units. From there, they flow at high pressure - 100 to 200 bar - into a mixing chamber. The flow rate ratio between the different streams must be precisely metered. In the mixer head, the streams impinge at high velocity, mix and begin to polymerise as they flow out in the mould cavity. Because the mixture is initially at low viscosity, low pressures – 2 to 10 barsuffice to fill the mould. The mixing chambers, the metering heads, the mixhead and the pumps take over the role performed by the plasticising unit in conventional injection moulding.

The reaction in the injected mixture generates some heat and also an increase in volume initially but as the polymerisation progresses and solidification sets in, contraction takes place. Some holding pressure may be needed to compensate the contraction but it can be bypassed by adding a small amount of a foaming agent, usually monofluor trichloromethane or triflouro trichloroethane, at the time of creating the mixture.

The moulds are maintained at a temperature between 40–65°C depending upon the composition of the polymer.

The manufacturing process takes place in the following sequence:

a. High pressure metering of reactants to the mixer head
b. Filling the mould (at high speed)
c. Curing by chemical reaction (polycondensation)
d. Demoulding
e. Finishing - flash trimming, post curing, cleaning, painting etc.

THE MOULDING MATERIALS

The RIM process has evolved from the PU rigid foam technology for manufacturing large articles which were difficult to mould in conventional process. Although the range of materials which can be processed by RIM has been extended, polyurethane in various forms such as flexible, semi-rigid, solid, foamed, reinforced and gas assisted constitutes over 90% of consumption. Other materials found suitable for the process are: nylon block copolymers, acrylics, polyureas, polyesters, epoxies, dicyclopentadiene (DCPD) and hybrid urethane systems.

Glass fibres are commonly used for reinforcement. They increase flextural strength and reduce shrinkage. Scrap plastics and wood flour too have been added for particular applications.

THE EQUIPMENT

The equipment for the RIM process consists of:

a. Conditioning system to prepare the ingredients for use
b. Metered pumping system to transfer intermediates for mixing
c. Mixing heads, where the liquid ingredients are combined through impingement
d. Mould carrier to hold, open, close and swivel the mould

Figure 16.1 depicts the first three steps of the process and Fig. 16.2 shows the mould carriers schematically.

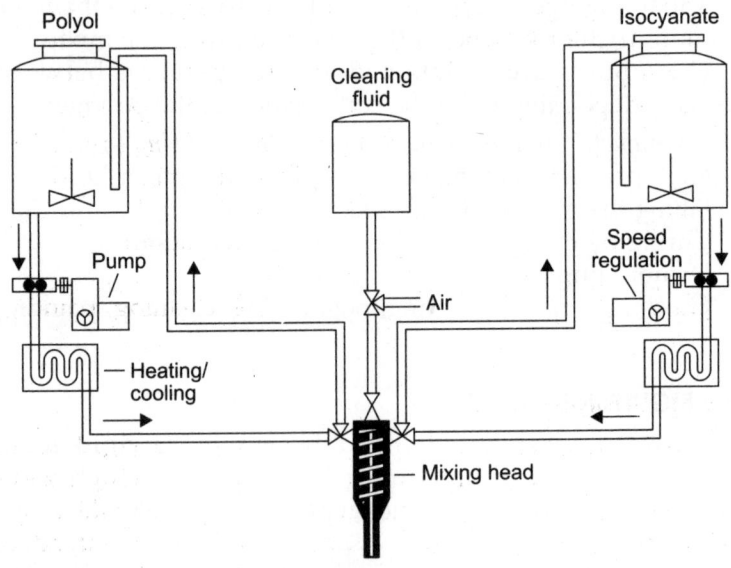

Fig. 16.1

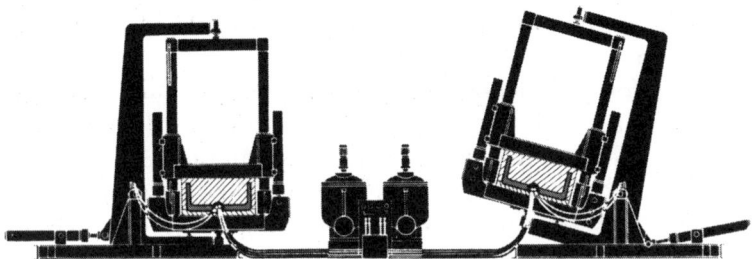

Fig. 16.2

MOULDS FOR RIM

The three characteristics, viz. the low viscosity. low pressure and low temperatures enable use of a wide range of mould construction materials, depending upon the quality, quantity and the cost of the products.

Silicone rubber moulds prove economical in cost and time when a few pieces are required for trials and prototypes. The quality of the surface is low.

Moulds, cast out of epoxy resins, render the surface details much better. They can be used to produce fairly large numbers of articles in foamed systems. The production rate is quite low.

Moulds made by spraying of low melting temperature tin-bismuth alloys can be produced very fast. The reproduction of the details of the model, too, is very good. The heat dissipation is also better than with resin moulds. The alloys have, however, a very low tensile strength and are bound to give rise to flash. The alloys are very expensive but can be used again and again.

Moulds for RIM can also be cast out of special zinc alloys like kirksite and aluminium alloys. Although more expensive, these offer higher dimensional accuracy, better reproduction of surface details, faster production and longer service life.

The optimum results are obtained by employing hardenable alloy steels or prehardened steels as mould materials. These can be polished to a very high degree which counters the tendency of urethanes to stick to metals and makes ejection easier. Nickel plating adds to the ease of demoulding. The surface details are reproduced exactly. An efficient cooling system can be incorporated. Because of higher strength and hardness, the mould service life is long and the flashing remains minimum.

Electroformed nickel shells, suitably backed and fortified, also offer a good alternative for fabricating efficient RIM moulds. The quality of the surface is optimum.

The moulds have to be maintained at a constant temperature (40–65°C, depending upon the composition being used) within a range of ± 2°C. Hence the RIM mould requires balanced cooling lines like the ones for injection moulding. The mould orientation should be such as to allow filling from the bottom of the mould cavity, allowing escape of air through a controlled vent on the top.

The feeding system of RIM moulds is similar to that in conventional moulding. The sprue, runner and gate act also as a transition zone and transform the turbulent flow from the mixing head into a laminar one.

The ejection of the soft PU-mouldings is carried out by stripping. Because of the very low injection pressure, the RIM equipment does not require high clamping forces, which may be as low as 2–5% of those needed for the same article if moulded in conventional injection moulding process. Moulds for small articles need not be clamped in a press; they can be locked by latches hinged with the mould halves.

APPLICATIONS

The RIM process has evolved from the PU rigid foam technology. The major application has been the facia of car bumpers in USA. Later, automobile fenders and body panels were made of glass filled urethanes. Though it is primarily being used for exterior automobile parts, there has also been significant application for interior car parts, appliances and recreational equipment. Over 90% of RIM production is polyurethane, as flexible, semi-rigid or micro cellular components. The advantage of RIM over conventional injection moulding includes the moulding of parts larger than 5 kg, those with very thin walls as the viscosity of the resin is very low and also those with very thick walls because curing is uniform throughout the moulding.

Some typical products made by RIM are:
- Large automotive parts like bumpers, spoilers, side panels
- Housings for electronic equipment, computers, copiers, type writers
- Housings, covers, large components for domestic appliances,
- Office furniture
- Outdoor furniture for beaches
- Bathroom articles like shower cabinets, tubs, cupboards
- Toilet seats, urinal partitions,
- Sports goods, snow-mobile parts
- Submersible sewage pump components
- Medical and laboratory equipment, dentists chair and wheel chair parts, etc.
- Lighting fixtures

ADVANTAGES

- Low cost tooling
- Design freedom; combination of thick and thin sections unproblematic
- Large L:T ratios possible due to low viscosity of components
- Excellent reproduction of surface details
- Flexibility in physical properties of mouldings through variation in material composition
- Higher strength with less weight by foaming

Bibliography

1. Adolf Franck. Kunststoff-Kompendium. Vogel Buchverlag, Wuerzburg, 2000.
2. Airmould. Battenfeld Gmbh. Germany and Austria, 2004.
3. AIRMOULD/AQUAMOULD Fluidunterstuetztes Spritzgiessen. Wittmann Battenfeld GmbH., Austria, 2008.
4. Batra. Der Faltkern. German Patent no. 10010611.2.24, 2002.
5. Batra. Design of Blow Moulds. CBS Publishers & Distributors, New Delhi, 2007.
6. Becker, Braun. Kunststoff Handbuch 1: Hanser Verlag, Munich.
7. C Gauthier, EH Heimann, E Nieksch, Revolution aus der Retorte, EHAPA-Verlag Stetten, 1966.
8. Catamold Feedstock for Metal Injection Moulding, Technical information, BASF, 200373.
9. CELLIDOR, Bayer AG, Leverkusen, 1960.
10. Charles A. Harper. Handbook of Plastics, Elastomers and Compounds. McGraw Hill, 2002.
11. Christoph Jaroschek. Spritzgiessen fuer Praktiker. Hanser Verlag, Munich, 2003.
12. Christoph Schumacher, Powder Injection Moulding for Novel Part Shapes, Kunststoffe 9/1999, Hanser Verlag, Munich.
13. CM Brockmann. Spritzpraegen technischer Thermoplaste, IKV Band 84, velag Mainz, 1999.
14. Colorants for Plastics. Popular Plastics and Packaging, Mumbai 12/2007.
15. Das Spritzgiessen von Akulon. Akzo Plastics bv, Zeist holland. 1975.
16. Dieter Rheinfeld. Verarbeitung vernetzbarer Formmassen in Spritzguss, Doctrate thesis, RWTH, Aachen, 1973.
17. Dipl. Ing. H.Eckardt. Multi Material Injection Molding Processes. Battenfeld GmbH. Meinerzhagen, 2003.
18. Dipl.Ing. H. Eckardt. Exploring the commercial benefits of water-assisted injection moulding. Battenfeld GmbH. Meinerzhagen, 2004.

19. Dr AK Kulshreshtha & Dr SK Awasthi. Trouble shooting in Polymer processing, Popular Plastics and Packaging, July 1997.
20. Dr Erwin Herrmann. Kunststoffeinfärbung, Zechner & Hütig Verlag GMBH, Speyer, 1976.
21. Dr Georg Schulz. Die Kunststoffe. Carl Hanser Verlag, Munich, 1959.
22. Dr Ing K Mienes. Kunststoffe in Amerika, Carl Hanser Verlag, Munich, 1954.
23. Dr W Michaeli. Einfuehrung in die Kunststoffverarbeitung. Carl Hanser Verlag, Munich, 2006.
24. Dr W Schoenthaler. Duroplaste. Technische Vereinigung, Wuerzburg, 1993.
25. Erich Gruber. Polymer chemie. Dr Steinkopff Verlag, Darmstadt, 1980.
26. F Johannaber, W Michaeli. Handbuch Spritzgiessen. Hanser Verlag, Munich, 2004.
27. Franck, Biederbick. Kunststoff-Kompendium, Vogel-Buchverlag, Wuerzburg, 1988.
28. Gas Injection moulding Technology-GIT, Processing Technology, ARBURG, 2004.
29. Gerd Poesch, Walter Michaeli. Injection Molding, Hanser Publishers, Munich 1995.
30. Gordon B. Thayer. Plastics Molds. Huebner Publictions, Cleveland, 1946.
31. Guide to Surface Defects, Kunststoff-Institut Luedenscheid, 2001.
32. H Dominghaus. Die Kunststoffe und Ihre Eigenschaften. Springer Verlag, Berlin, 1998.
33. H Wolfram, M Bloemacher, D Weinand. Powder Injection Moulding of Stainless Steels, BASF, 1998.
34. Handbuch der Temperaturregelung mittels flüssiger Medien. Regloplas AG, St. Gallen, 1986.
35. Hans Stoeckert. Die Kunststoffe und ihre Eigenschaften. VDI Verlag, Dusseldorf, 1976.
36. Hans-Georg Elias. An Introduction to Plastics, Wiley VcH, Weinheim, 2003.
37. HASCO-Normalien D1 & D2. HASCO Hasenclever GmbH & Co. KG. Germany.
38. Heisskanaltechnik. HASCO Hasenclever GmbH & Co. KG. Germany 2007.
39. Herbert Rees. Mold Engineering. Hanser Publishers, Munich, Vienna, New York, 2002.

40. Hermann V. Boenig. Structure and Properties of Polymers. George Thieme Publishers, Stuttgart, 1973.
41. Hotrunner System, XINTECH Systems, Duebendorf, Switzerland, 1997.
42. Ian M Campbell. Introduction to Synthetic Polymers. Oxford Science Publications, 1994.
43. "Injection Moulding India". Society of Plastics Engineers (Indian Section), 1997.
44. Injection moulding of liquid silicone rubbers, ARBURG GmbH & Co. KG, Germany.
45. Injection Moulding of Silastic SLR, Dow Corning Corportion.
46. Injection Moulding, Bayer Anwendungstechnik, Leverkusen, 1979.
47. Innovative Injection Moulding. Wittmann Battenfeld GmbH., Austria, 2008.
48. Irvin I Rubin. Injection Moulding, Theory and Practice. John Wiley and Sons, New York, 1976.
49. James F Stevenson, Innovations in Polymer Processing, Molding. Hanser Publishers Munich, Vienna, New York, 1996.
50. Jay Shoemaker. Moldflow Design Guide. Hanser Verlag, Munich, 2006.
51. JH Dubois, FW John. Plastics van Nostrand Reinhold Co. New York, 1981.
52. John Brydson. Plastics Materials, Butterworth, Heinemann, Oxford, 2000.
53. Joseph B. Dym. Injection Molds and Molding. Van Nostrand Reinhold Compny, New York, 2003.
54. KF Hens, C Bader, Pulver-Spritzgiessen. Kunststoffe 4/1995, Hanser verlag, Munich.
55. Klaus Stoeckhert. Mold-Making Handbook, SPE, Hanser Publishers, New York.
56. KM Reinfrank. Die Spritzgussverarbeitung thermoplastischer Formmassen. BASF, Ludwigshafen, 1989.
57. Kunststoff-Physik im Gespraech, BASF AG, Ludwigshafen.
58. Kunststoff-Verarbeitung im Gespraech-1, BASF AG. Ludwigshafen, 1987.
59. Kunststoff-Werkstoffe im Gespraech, BASF AG, Ludwigshafen.
60. L Lindner, P Unger. Gastrow-Injection Moulds: 100 Proven designs. Hanser Verlag, Munich, 1993.
61. Marc Ros, GE Plastics. ICM's Winning Ways, Asian Plastics News 3/2000.

62. Menges, Haberstroh, Michaeli, Schmachtenberg, Werkstoffkunde Kunststoffe. Carl Hanser verlag, Munich, 2002.

63. Menges, Mohren. How to Make Injection Moulds. Hanser Publications, New York, 1993.

64. N Mehta, B Singh. Mumbai Institute of technology. "Gas Assisted Injection Moulding", Popular Plastics and Packaging, Mumbai. June, 2002.

65. Otto Schwarz. Kunststoffkunde. Vogel Buchverlag, Wuerzburg, 1988.

66. Peter Eyerer, Peter Elsner, Marc Knoblauch-Xander, Andreas von Riewel. Gasinjektionstechnik. Hanser Verlag Munich, 2003.

67. Peter Jones. The Mould Design Guide. Smithers Rapra Technology Ltd. Shawbury, UK 2008.

68. Peter Unger. Gastrow-Der Spritzgiesswerkzeugbau. Hanser Verlag, Munich, 2007.

69. Piechota, Roehr. Integralschaumstoffe, Carl Hanser Verlag, Munich, 1975.

70. Powder Injection Moulding (PIM), Technical Information, ARBURG Gmbh + Co. KG Germany.

71. Processing Elastosil SR, Liquid Silicone Rubber, Wacker-Chemie GmbH. Munich.

72. Rainer Protte, Klaus Kronejung. Wasserinjektionstechnik-WIT mit Bayer Polymeren. Bayer Polymers, 2003.

73. Rajesh R. Wadhwa. Two platen injection moulding machines. Popular Plastics and Packaging, Mumbai, 3, 2007.

74. Ralph E Wright. Injection/Transfer Molding of Thermosetting Plastics. Hanser/Gardner Publications, Cincinnati, 1995.

75. Raymond Foad. Gas Injection Moulding Technology. Cinpress Ltd. UK, 1997.

76. RGW Pye. Injection Mould Design, Longman Scientific & technical, Essex, 1989.

77. RP Swool, XS Sun. Bio-Based Polymers and Composites. Elsevier Academic Press, UK USA, 2005.

78. S Levy-JH Dubois. Plastics Product Design Engineering Handbook. Van Nostrand Reinhold Company, New York, 1977.

79. Sasan Habibi-Naini: Neue Verfahren fuer das Thermoplastspritzgiessen, IKV Aachen Band 155, 2003.

80. Schwarz, Ebeling, Furth. Kunststoffverarbeitung. Vogel Buchverlag, Wuerzburg, 1999.

81. Spritzgiessen von Thermoplasten, Farbwerke Hoechst AG, Frankfurt (Main) 1971.

82. Spritzpraegen, Process Tecnology. Demag Plastics Group, Wiehe, Germany.
83. Stitz, Keller. Spritzgiesstechnik. Hanse Verlag, Munich, 2001.
84. Tadmor, Gogos. Principles of Polymer Processing. John Wiley and Sons, Hobokon, 2006.
85. TC Pearson-Gas injection Ltd. Benefits and Limitations of GAM, Popular Plastics and Packaging 8/1998.
86. Technical Manual IXEF, Solvay Advanced Polymers, Solvay SA, Brussels, 2001.
87. Terry Pearson. Innovatory Applications Using Gas Assisted Moulding for the Automotive and Consumer Durable Industry. Gas Injection Ltd. UK, 1997.
88. Terry Pearson, RR Wadhwa. Latest Innovations in AIM. Popular Plastics and Packaging, Mumbai, February, 2004.
89. Uwe Wolf. Typical Injection Moulding Problems and How to avoid Them. BASF, 1989.
90. VE Yarsley. Plastics Applied. The National Trade Press, London, 1945.
91. Verfahrenstechnische Alternativen und Verfahrensauswahl. ATI 1147d. Bayer AG, Leverkusen, 2002.
92. Warpage Design Principles. Colin Austin, Moldflow Pty. Ltd. Victoria.
93. Wilhelm Keil, Kunststoffe. Wiley-VCH verlag, Weilheim, 2006.
94. Willi Dalhoff. Verarbeitung von duroplastischen Formmassen durch Spritzgiessen, Doctrate thesis, 1969.
95. Wirtschaftliche Fuehrung eines Spritzgussbetriebes, VDI-Verlag, Dusseldorf, 1976.
96. Wolfram Taenzer. Biologisch Abbaubare Polymere, DVG Stuttgart, 2000.
97. YS Vani. Multi-Material Moulding Processes, Part I & II, Popular Plastics and Packaging, May and June, 2006.

Index

389